Windpower Ownership in Sweden

Windpower is a modular technology and, compared to most other power plants, also rather small scale. Wind turbines are installed to the distribution grid, which is called *distributed generation*. Ownership and revenues can be distributed as well, using the right ownership models.

Several different ownership models used in different countries, such as windpower cooperatives, local limited companies, net-accounting models, and the like, are described. These models can be used to make windpower become community power and/or consumer-owned power and produce power for these owners at cost price.

To wait for the international community to agree on international treaties to create a renewable energy system takes too long. The climate disaster will inevitably happen. To speed up the transition to a renewable energy system, development has to come from below, from local communities that can initiate and invest in windpower and other renewables. At the same time, to get off the oligopolistic electric power market and produce power at cost price is good business.

Tore Wizelius is a writer who has authored eight books on windpower. He was a lecturer on windpower at Gotland University from 1998 to 2008 and is now working as a windpower project developer in Sweden.

Windpower Ownership in Sweden

Business models and motives

Tore Wizelius

Routledge
Taylor & Francis Group
LONDON AND NEW YORK

First edition published 2013
by Media-Tryck, Lund University
Lund, Sweden

This edition published 2014
by Routledge
2 Park Square, Milton Park, Abingdon, Oxon OX14 4RN

and by Routledge
711 Third Avenue, New York, NY 10017

Routledge is an imprint of the Taylor & Francis Group, an informa business

British Library Cataloguing in Publication Data
A catalogue record for this book is available from the British Library

Library of Congress Cataloging in Publication Data
Wizelius, Tore, 1947–
Windpower ownership in Sweden : business models and motives /
Tore Wizelius.
pages cm
Includes bibliographical references and index.
1. Wind power – Sweden. I. Title.
TJ825.W592 2014
333.9′209485–dc23 2013043317

ISBN: 978-1-138-02111-2 (hbk)
ISBN: 978-1-315-77798-6 (ebk)

Typeset in Sabon
by HWA Text and Data Management, London

Contents

Illustrations

Figures

Tables

Acknowledgements

I express my very great appreciation to my research supervisors, assistant professor Jamil Khan and professor Lars J. Nilsson, for valuable and constructive suggestions during the planning and development of the research work and their guidance, encouragement, and useful critiques during the process of writing this report. I am also grateful to Gotland University for financing this study.

I am particularly grateful for the assistance given by Carin Persson for the help with personal interviews with several informants. She did excellent work. I also thank all informants who have given their time to explain their business models and motives.

Special thanks to Thomas and Peter Sirland who, during the years, have provided me with very valuable information on the business side of windpower development. I offer my special thanks to Nils Erik Carlstedt, who gave me access to the wind turbine database managed by Vindstat AB.

Finally, I thank Annika Gustavsson for her support and encouragement throughout my study.

Executive summary

The transition to a renewable energy system in Sweden by development of windpower plants has been conducted by non-utility generators and some municipal utilities. Investments have been made by private persons, farmers, and small and medium-size enterprises, either directly or by shares in windpower limited companies or cooperative associations. Project developers have designed ownership and business models for these different owner categories to make it possible to finance windpower plants.

Recently, some large industrial companies have joined this development by forming joint ventures with transnational power companies. Also public and private companies in other lines of business, such as real estate companies, have invested in windpower plants to secure stable and predictable long-term prices for their electric power consumption.

Several municipal utilities have played an important role in the development of windpower in Sweden. These have had political ambitions and have taken responsibility for the transition to a renewable energy system that has been set as a target by the Swedish parliament. These municipal utilities have also promoted windpower cooperatives and have given small and medium-size companies opportunities to become windpower owners.

Since windpower development in Sweden started in the early 1990s, the share of different owner categories has varied. The share of windpower limited companies has doubled. Power companies and utilities had a large share in the very beginning, but their share has decreased to around 20 percent. The share of private and cooperative ownership is declining. Companies in other lines of business have had a share just below 20 percent during all this period from 1990 to 2012.

The share of power companies and utilities of the windpower capacity in Sweden in 2012 was 24 percent. If power companies and utilities are separated, it turns out that the share of power companies has declined from 22 percent in 2007 to 11 percent in 2012 but that the share of municipal utilities during the same period has increased from 6 percent to 13 percent. The share of "community power" (i.e., power plants with local/regional ownership) is estimated to be at least 40 percent and the share of "consumer power" around 6 percent.

The share of the state-owned power company Vattenfall, which owns one-third of the power companies' windpower capacity (less than 7 percent of the total capacity), is declining. This is surprising as Vattenfall, according to the directives from the government in 2009, should be one of the companies that *lead the development of an environmentally sustainable energy production.*

The share of windpower owned by municipal utilities and companies has been increasing. This indicates that local politicians take more responsibility for the transition to a renewable energy system than the government, which is responsible for the policy of Vattenfall.

Many different business models are used. These models have been continuously developed and adapted for different kinds of investors. When the size of windpower plants and investment costs increased, different models for mixed ownership were developed. The use of such "umbrella companies" made it possible for these different ownership categories to continue to invest in windpower plants. Project developers have been very innovative and creative in this development of new models.

In the early 1990s, an investment in windpower was a "political statement," a demonstration of the will and possibility to create a sustainable electric power system. This was the motive of members of cooperatives, some local wind limited companies, but also of municipal utilities that developed windpower plants. This is still a common motive and is used by several companies for marketing windpower.

A strong motive for companies in other lines of business is to hedge power prices (i.e., to reduce the insecurity of unpredictable and volatile power prices on the market). This is a long-term strategic interest that motivates large investments in windpower plants. Some have additional motives: to strengthen their trade mark or utilize land areas for new purposes.

Private investors and municipal utilities are the main actors who implement a transition to a renewable energy system in Sweden by development of windpower. The conditions for these ownership categories to continue to play such a role should be secured and improved.

Since 2012, the revenues from windpower plants have been too low to make ends meet for those who made their investments a few years earlier. The owner categories that have driven the development of windpower in Sweden will no longer be able to finance windpower projects with such high risks.

The economic risk can be reduced by a *guaranteed minimum price* for certificates. Otherwise, investments by the ownership categories that develop windpower in Sweden will stop. Ambiguous rules and regulations are also an investment risk. These rules should be clarified by politicians in a *Windpower law* such as they have in Denmark.

1 Introduction

There is a broad understanding that a transition from a power system that relies heavily on fossil fuels to a renewable energy system based on renewable energy sources is necessary. This ambition is expressed on the international level as well as by national governments, nongovernmental organizations, and the general public. The European Commission has, for example, in 2007 set four targets and objectives for the year 2020:

- reduction of greenhouse gas emissions from developed countries by 30 percent;
- improving energy efficiency by 20 percent;
- raising the share of renewable energy by 20 percent; and
- increasing bio fuels in transport to 10 percent.

Windpower is the major and most cost-efficient renewable energy technology and is expected to provide the bulk of new renewable energy generation.

Windpower development started in the early 1980s and took off in countries such as Denmark and Germany in the mid-1990s. During the same time period, starting in the mid-1990s, there has been a parallel development toward a deregulation of the electric power markets. This has been part of a neo-liberal agenda (Högselius & Kaijser, 2007). The basic principle has been to unbundle the electric power systems so that power production, power distribution, and power trade are strictly separated. Free access to the grid is another important ingredient in this policy.

Is the transition to a renewable energy system made easier and more efficient by a deregulation of the electric power market? Has it caused or stimulated any changes in the ownership structures? Has it caused any other unpredicted effects?

If so, are these changes in ownership structures disruptive of the dominant energy system? Will the electric power system disintegrate into small-scale distributed generation, where each building will be self-sufficient with energy? Or will this new renewable technology be absorbed and integrated by the power-industrial complex?[1]

As there is a broad and global consensus that a transition to a renewable energy system is necessary and very urgent, it is also important to find out which means and factors stimulate a *fast* transition. If some ownership models are more efficient, this is an important message to decision makers, who then could adapt promotion strategies, rules, and regulations to enable these models to be used.

From an environmental point of view (i.e., to reduce greenhouse gas (GHG) emissions as fast as possible), it makes no difference whether this is made by large transnational power companies or by local cooperative associations, private households, or industrial corporations. There may, however, be other advantages or disadvantages with different ownership structures.

The global energy system is very complex and diverse; so is the electric power system. Here the focus will be on one renewable electric power technology: windpower. This will be a case study within this wider context. The analysis of ownership models will focus on models used in Sweden, but some examples from Denmark, the United States, and some other countries are described as well.

The deregulation of the electric power market has created opportunities and openings for innovative companies, entrepreneurs, and individuals to find new solutions for power production, distribution, and sale. New business models and corporate strategies have also appeared. The policy measures to promote renewable energy sources have created new financial instruments that also affect market prices and development of windpower and other renewable energy sources.

Several new trends can be identified. Many of these have been discernible after the turn of the century.

1.1 Own your own power plant

One of these trends is that power consumers produce their own electric power with their own power plants. This trend manifests itself on different levels.

1.1.1 Individual

There are individual solutions where private consumers develop "moonshine" power by installing stand-alone systems with wind turbines, photovoltaic cells, batteries, and the like on their own premises, without any connection to the grid or with a grid connection and so-called net metering (where the meter turns backward when power is fed to the grid). This will make them independent of the power market and self-sufficient with electric power.

1.1.2 Cooperative

Another expression of this trend is cooperative ownership, where some hundred power consumers—households, farmers, and/or companies—own and operate their own power plant (in most cases, wind turbines) with shares that correspond to their individual power consumption.

1.1.3 Corporation

Lately, also large corporations have joined this trend, such as the Swedish forestry company SCA, that now is developing windpower plants[2] on their own premises (forests, harbors, etc.) that will produce power for their factories and/or for sale on the market.[3] Many municipal and private housing companies develop or buy windpower plants that will produce electric power for their houses and tenants.

This trend for consumers to produce their own electric power can be interpreted as an *escape from the market*. The availability and the cost for electric power will be predictable and defined by the cost of production instead of by supply and demand on the electric power market.

1.2 Aim and research questions

The purpose of this investigation is to find and describe the legal, organizational, and economic preconditions for this practice and trend toward user ownership of power plants. Another aim is to find the motives these actors have for joining this development.

What are the advantages/disadvantages with these ownership models for a fast transition to a renewable energy system; or, more specifically, for a fast development of windpower in Sweden? To find answers to these questions and to analyze this dynamic development and the repercussions for the power market are other aims of this study.

The final aim is to write a thesis that is interesting and *readable*, not only for colleagues in the academic world but for the interested public, just as I have tried to do with the textbooks on windpower that I have written in the past. I will try to explain complicated things in an understandable way. New knowledge has to reach the public to become common knowledge. Hopefully, this will increase the speed of a transition to a renewable energy system.

1.2.1 Research questions

These aims can be transformed to the following research questions:

1. Which ownership models for windpower are used in Sweden, and how are they designed?
2. How have these ownership models and their share of the total installed windpower changed over time?

These two questions are closely connected and the result of a quantitative analysis and a descriptive presentation of the models used.

3 Why are these ownership models used?
4 What are the advantages/disadvantages with the ownership models for a fast transition to a renewable energy system; or more specifically, for a fast development of windpower in Sweden?
5 How can laws, rules, and regulations be adapted to promote these ownership models, given that they have advantages?

1.2.2 *Scope and method*

The scope of this study is developments within the sector grid-connected wind turbines, with some historical background and an outlook into the future.

The main focus will be ownership models used in Sweden for grid-connected wind turbines. Some international comparisons will be made, but the focus and most of the facts and figures will be from Sweden. This can be regarded as a case study on Sweden in a global context.

Within each of these options to own your own power plant mentioned in Section 1.1—the private, the cooperative, and the corporate—a dynamic development can be discerned. However, in this context I will focus on cooperative and corporate ownership and leave the individual solutions outside the scope of this thesis.

The methods that are used are collection and analysis of facts and figures from official documents, media, and interviews. Information on aims, motives, and business ideas have been collected by open and semi-structured interviews with persons who are representing different owner categories (see Appendix A). Some of these facts, figures, and information from interviews are reused from work done with three windpower manuals: one for farmers, a second for cooperatives, and a third for municipalities, made for the Swedish Energy Agency in 2009–2010 (Wizelius, 2009, 2010/2012a, 2010/2012b).

For some windpower owners, mainly power companies and some utilities, the motives have been taken from policy statements and so on from their Websites. The purpose of some information on company Websites is for marketing the company and its products. Companies may have motives other than those that are published on public Websites. For windpower owners who have not been interviewed, only these public motives have been cited.

The environmental challenges of today are caused by the interaction between nature, humans, society, and technology. To handle such questions and problems it is necessary to use *interdisciplinary* methods. Facts and figures are quantitative data and are presented in statistical diagrams and the like. To analyze the electric power market, it is necessary to use and discuss economic theory. Interviews give qualitative data and are analyzed and interpreted with heuristic methods (Öberg, 2008).

The theoretical framework is based on the socio-technical system concept initially developed by Thomas P. Hughes (Hughes, 1993). Theory will be discussed further in Chapter 2.

1.3 Outline

In Chapter 2, Transition of Socio-Technical Systems, there is a short overview of relevant theory on socio-technical systems. This overview will introduce the concepts used in the analysis, a theoretical framework, and the wider context for this study.

In Chapter 3, Windpower in the Electric Power System, the technical, economic, and political aspects of this system are described. A proper understanding of this is precondition for understanding the options for a transition to a renewable energy system and the role of ownership in this context.

Chapter 4, Windpower Development and Ownership, describes the development of windpower technology and ownership on a global level, with examples from a number of countries, to give a background and context to the analysis of windpower ownership models used in Sweden.

Chapter 5, Ownership Models in Sweden, describes and compares ownership models used by different actors in Sweden.

Chapter 6, Windpower Ownership in Sweden, presents facts and figures on ownership models used in Sweden. The ownership models are finally discussed and analyzed one by one.

Chapter 7, Windpower Owners' Motives, investigates why these different ownership models have been chosen and describes the motives of different owner categories to develop, invest in, and/or own and operate windpower plants.

In Chapter 8, The Relevance of Windpower Ownership, the results of the investigation are summarized and put in a wider context by discussing how the findings fit the theories presented in Chapters 2, 3, and 4. The two remaining research questions, on advantages/disadvantages of these ownership models and how laws, rules, and regulations can be adapted to promote these ownership models, are discussed.

The final chapter, Chapter 9, Future Outlook, draws conclusions, and some ideas for future research are presented.

1.4 Personal background

I'm writing this thesis not at the start but at the end of my professional career. I planned to get it ready the same year as I retired, according to Swedish law (i.e., when I turned 65). However, owing to my work as a windpower project developer, a time-consuming and tedious job, it took another year.

Although I'm of mature age, my academic career has been very short. In 1997, I suggested to the Extension Centre of the Royal Technical University in Visby, Gotland, that they should develop some courses on windpower. They then asked me whether I could do that, and I accepted. The Extension Centre was then integrated into the newly established Gotland University where I stayed for some ten years, developing courses, mainly for distance education, first in Swedish and then also in English, and teaching these courses as well.

It started out as a side job, but in 2008, there were ten teachers employed to manage and teach some ten different courses on windpower, focused not on technology but on windpower project development, which by nature is an interdisciplinary topic, comprising technology, meteorology, economics, politics, law, management, and communication.

As there was no university textbook on windpower and my students needed one, I wrote the book *Vindkraft i teori och praktik* (Wizelius, 2003/2007) in 2003, now printed in two editions and a third one to be published in 2014. The book was then translated and published in English in 2007, *Developing Wind Power Projects* (Wizelius, 2007), with a second edition to be published in 2014 as well. In 2011, another textbook, *Wind Power Plants & Project Development*, which I wrote together with my Indian colleague, professor Joshua Earnest, was published by PHI Learning in India (Earnest & Wizelius, 2011). Finally, I have contributed with a chapter on *Design and Implementation of a Wind Power Project* in the reference book *Comprehensive Renewable Energy* published by Elsevier in 2012 (Wizelius, 2012).

My very first book, *Ack, Värmeland du sköna—avfolkning och bolagsvälde i norra Värmland*, was published in 1972. In English, the title would be Oh, beautiful Värmland—Depopulation and Corporate Power in Northern Värmland (Jacobsson & Wizelius, 1972). This was originally a paper corresponding to a master's thesis (there were no master programs at Swedish universities at that time) in social anthropology.

In retrospect, I realized that the approach in this first book actually was very similar to the one I use in this thesis: a study of a socio-technical system, forestry, with ownership models as the important parameter (a comparison of local communities with forests owned by farmers versus forests owned by forestry companies), as well as the impact of new technology for mechanization of forestry on the ecological system, social structure, economy, and local politics.

After finishing my studies at Stockholm University, I have been working as a journalist, most of the time as a freelance writer. As I have been a devoted environmentalist since the 1960s, I specialized in topics related to environment and energy. I was engaged in the anti-nuclear movement, convinced that the risks of nuclear power were too high, which unfortunately has been confirmed by the accidents in Harrisburg, Chernobyl, and Fukushima.

However, to close down nuclear power plants and replace GHG-emitting fossil fuel-fired power plants, there has to be renewable energy sources that can replace them. I found it more rewarding to work for and promote solutions than to criticize the energy policy of the government. Windpower is a renewable energy source, and I followed and wrote articles and reports on the development of windpower. As a spin-off, I became an expert on this subject. I understood that when I made an interview with an engineer in charge of windpower at a very large Swedish power company and realized that he didn't have even a basic understanding of windpower technology or wind meteorology.

In the 1990s, I was correspondent in Sweden for the international magazine *Windpower Monthly*. In the early 1990s, I wrote my first books on windpower, *Vind, del 1, Vind, del 2,* and *Vind, del 3* (Karlsson & Wizelius, 1992; Wizelius 1993, 1994). The idea was to have a yearbook on windpower development, but after three books, the publisher gave up the idea. My fourth book on windpower, *Vindkraft—en ny folkrörelse*, (Windpower—a new people's movement), was about windpower cooperatives and was published in 1999 (Wizelius, 1999).

My interest in cooperatives also had a background. In the early 1980s, my family became members of a nursery school cooperative in Stockholm. This was a rather new social invention at that time, there were only four such cooperatives in Sweden. Local politicians were not so fond of this, not until they realized that such cooperative childcare operated by parents was in fact much cheaper for society than public nursery schools. Politics changed quickly, and today there are several nursery co-ops operating in all Swedish municipalities, and all across Sweden there are thousands.

While I still was a member of one, when my kids still were young, I wrote a manual on how to start such cooperatives, which helped other parents to follow in our footsteps. And there was a chain reaction, leading up to the thousands of parent-operated cooperatives that exist today. By writing this manual for windpower cooperatives, I hoped to start a similar social chain reaction.

During the 1990s I was also board member and secretary in the Swedish Windpower Association for several years as well as editor for the association newsletter. As soon as I started to work at Gotland University, I left these assignments, as a university teacher should be unbiased and not be a member of any lobby groups.

After taking leave from my job at Gotland University in 2008, I have devoted half of my time to prepare for and work on this licentiate thesis financed by my employer, who judged that I would need some academic credentials after retirement. At this university, by far the smallest in Sweden, employees do not get credit for what they do or can do, or for textbooks, but only for strictly academic credentials (i.e., formal rather than actual competence). As a consequence, Gotland University ceased to exist in July 2013, when it returned to be an extension center, this time for Uppsala University.

I had the distressing experience to see a new, very progressive and creative workplace turn into a conservative medieval institution, based on academic caste membership, in less than ten years. I was quite happy to leave and get the opportunity to move to a more mature academic culture at Lund University. In 2010, I also developed a master-course on *Windpower*

Planning for landscape architects at the Swedish University of Agriculture at Alnarp outside Malmö.

In the other half of my time I have written some new books (Wizelius, 2009/2010, 2010/2012a, 2010/2012b; Earnest & Wizelius, 2011), revived my private company Vindform AB, (windpower consultancy and publishing house), and developed a package of distance education courses on windpower translated into Chinese and used for competence building in China, financed by the German technical cooperation agency GTZ. I was appointed board member in a windpower investment fund (WindCap) and also started a number of windpower project development companies, such as Boge Vindbruk AB and ReWind Offshore AB, developing near-shore offshore wind farms with mixed public and private ownership.

I have also participated in several different projects within the windpower network (Nätverket för vindbruk, 2012) managed by the Swedish Energy Agency. I initiated and participated in a workgroup, with representatives for project developers, the farmers' organization LRF, power companies, and other interested parties, which developed a common recommendation for land lease contracts for windpower.

After that, another workgroup was formed to develop similar recommendations for local benefits (i.e., agreements made with local communities guaranteeing that a reasonable part of the economic benefits would stay at the local level). Further, I participated in a number of regional seminars informing municipal companies and politicians on the opportunities and advantages of investing in their own windpower plants.

By now, I thus have an extensive network in the windpower branch in Sweden and also internationally. Many persons in windpower developing companies, public administration, and others that I needed to interview for this thesis I have known personally for years, and some have been my students. This creates a problem, as it is difficult to have unbiased interviews with friends. To solve this, I engaged an assistant, Carin Persson, to interview these persons and record them on tape. The interviews I have made myself are also recorded and transcribed to paper.

All these contacts and the engagement in companies in the windpower business have, however, also given me very valuable inside information and knowledge on business models and project evaluation. Business information is confidential, but it has given me a good general background knowledge that may be useful also in this context.

Obviously, with this personal background, I ought to be biased in favor of windpower and of cooperatives, since I have promoted this technology and this ownership model. However, my background as a journalist, with strict professional rules on objectivity, and at the university, with similar rules, makes me quite conscious about this risk and also aware of the necessity to have a critical attitude to the topic and to my own personal evaluations. I hope that my background knowledge and experience will in fact help me to analyze the topics of this thesis in an unbiased and objective way.

Notes

1 The term *power-industrial complex* has the same meaning as the military-industrial complex, an expression first used by President Eisenhower. It denotes the group of very large transnational power companies and power equipment companies that have formed the dominant energy system (DES).
2 A group of wind turbines are often called *wind farms* or *wind parks*. I call one or a group of wind turbines a windpower plant, as they are power plants driven by the wind and produce electric power, just like any other power plant.
3 Both options give companies control of future costs for electric power.

2 Transition of socio-technical systems

This chapter gives a short overview of relevant theory on socio-technical systems and presents some ideas on how these can be used and further developed in this context.

2.1 Socio-technical systems

The concept *socio-technical systems* was developed by Thomas P. Hughes in his book, *Networks of Power; Electrification in Western Society, 1880–1930* (Hughes, 1993). The basic thesis is that a technological system cannot be understood as a strictly technical system but that such technical systems develop in a process that involves several other factors, such as politics, economics, legislation, and so on.

Hughes starts his study in 1880, when the first electric power systems were installed, and ends in 1930, when the system had reached a mature stage, when no more radical innovations could be expected and the electric power systems had become integrated in national transmission grids.

This concept has since Hughes's path-breaking book been further developed and is discussed in the book *Social Construction of Technological Systems* (Bijker et al., 1989) edited by Wiebe E. Bijker, Thomas P. Hughes, and Trevor Pinch.

2.1.1 Development of socio-technical systems

In *Networks of Power* (Hughes, 1993), Thomas P. Hughes identifies several stages in the development of electric power systems:

1 Invention and development: Inventor-entrepreneurs play a dominant role during this stage. Engineers, managers, and financiers are also involved but do not preside over the system's growth.
2 Technology transfer from one region and society to another: The agents of change include inventors, entrepreneurs, organizers of enterprises, and financiers.

3 System growth: The key concept here is "reverse salients" (i.e., critical problems that appear and prevent growth and the solutions that make further growth possible). If these problems cannot be solved, occasionally new systems may emerge.
4 Momentum: As the system grows it acquires momentum – mass, direction, and velocity.

Hughes started out as an electrical engineer and then became an academic historian. Thus he uses these metaphors from physics: mass, momentum, velocity, and so on. He characterizes this fourth stage in the development of the electric power system, the phase that this system is in today, in this way:

> In the case of technological systems, as defined in this study, the mass consists of machines, devices, structures and other physical artifacts in which considerable capital has been invested. The momentum also arises from the involvement of persons whose professional skills are particularly applicable to the system. Business concerns, government agencies, professional societies, educational institutions, and other organizations that shape and are shaped by the technical core of the system also add to the momentum. Taken together, the organizations involved in the system can be spoken of as the system's culture.
>
> (Hughes 1993, p. 15)

In his book, Hughes describes in detail the evolution of the electric power system in three major cities: Berlin, Chicago, and London. In this context, he also describes and discusses ownership models, which are further discussed in the section on Ownership in Chapter 4.

Finally, Hughes describes and compares three such regional power systems: *Reinisch Westfälisches Elektricitetswerk* in Germany, *Pennsylvania Power & Light Company* in the United States, and *Newcastle upon Tyne Electric Supply Company* in the United Kingdom. He finds that although the pool of technology available was the same, the entrepreneurial character as well as the power systems in these three regions differed owing to variations in geographical, cultural, managerial, and entrepreneurial character (i.e., the socio-political context).

Hughes's book ends in 1930, when the electric power systems had become mature, standardized, and a well-integrated technical system in most countries.

2.1.2 *System components*

In *The Evolution of Large Technological Systems* (Hughes, 1989), Hughes develops the concepts from his book *Networks of Power.* He starts out by defining technological systems, which he considers to be both socially constructed and society shaping. The components of a technological system are

- physical artifacts—generators, transmission lines;
- organizations—manufacturing firms, utilities, banks;
- science—books, articles, teaching programs;
- legislation—laws, regulations; and
- natural resources—coal, oil, gas, urnaium, wind, sun.

Artifacts that are components in a system interact with other artifacts. Artifacts and components in a system are mutually interdependent; if one changes, the others have to change as well to adapt to the *system demand*.

The system builders (described in *Networks of Power*) had the ability to force unity from diversity, centralization in the face of pluralism, and coherence from chaos. This construction often involves the destruction of alternative systems.

Components in the system interact; thus the management structure of a power company depends on the character of the artifacts working in the system and the other way round, and management often chooses technical components that support the structure of management. The technical system also interacts with the business strategy. These analogous structures and strategies make up the technological system and contribute to its *style*.

Hughes also observes that

> Over time, technological systems manage increasingly to incorporate environment (defined as "intractable factors not under the control of the system managers"), into the system, thereby eliminating sources of uncertainty, such as a once free market.
>
> (Hughes, 1989)

Inventions can be radical (they inaugurate new systems and do not become components in existing systems) or conservative (they improve or expand existing systems). Well-established "mature" organizations rarely nurture radical inventions.

Development

If radical inventions are successfully developed, they culminate in technological systems. Innovation is defined as bringing an invented product into use, in a complex system of manufacturing, sales, and service facilities. System builders strive to increase the size of the system under their control.

Technology transfer

This can occur at any time during the history of a technological system. Legislation and market are critical factors in transfer and adaption, and organizational components are transferred as well.

Technological style

Adaption is a response to different environments, and adaption to environment culminates in *style*. Why do electric light and power systems differ from time to time, from region to region, and from nation to nation? Possible explanations are *regulatory legislation*, *natural geography*, and *historical experiences*.

2.1.3 *Growth, competition, and consolidation*

The main cause of growth of the electric power system is that the cost efficiency increases with diversity of loads (i.e., it is inherent in the technical system and the consumption pattern). When the size of the system increases, the load factor[1] of power plants and thus the economic returns will increase as well.

During growth, *reverse salients* appear (i.e., components that have fallen behind), which cause problems that have to be solved. It can be described as friction in the system. Financing, organization, and the like can be such reverse salients. When a reverse salient cannot be corrected within the context of an existing system, the problem becomes a radical one, the solution of which may bring a new and competing system.

Hughes then refers to Edward Constant, who talks about *presumptive anomalies*, which occur when

> ...under some future conditions the conventional system will fail, or function badly, or that a radically different system will do a much better job.
> (Constant, 1989)

Mergers

Consolidation occurred until the interwar period, when two large manufacturers in the United States, *General Electric* and *Westinghouse*, and two in Germany, *AEG* and *Siemens*, dominated electric manufacturing. Large regional utilities prevailed in electrical supply. Industry-wide standardization of technical hardware was created, with standard voltage, frequency, and so on.

Momentum

Technological systems, even after long growth and consolidation, do not become autonomous; they acquire momentum. They have a mass of technical and organizational components; they possess direction or goals; and they display a rate of growth suggesting velocity. This is a quality that is analogous to the inertia of motion.

Manufacturing corporations, public and private utilities, industrial and government research laboratories, investment and banking houses, sections of technical and scientific societies, departments in educational institutions, and regulatory bodies add greatly to momentum of modern electric light and power systems.

Inventors, engineers, scientists, managers, owners, investors, financiers, civil servants, and politicians often have *vested interests* in the growth and durability of a system. Concepts related to momentum include *vested interests*, *fixed assets*, and *sunk costs*. Durable physical artifacts project into the future the socially constructed characteristics acquired in the past when they were designed.

2.1.4 The electric power system

As wind turbines produce electric power, it may be easier to first consider its relation to the electric power system, where wind turbines are relatively new technical artifacts and windpower can be regarded as a new component. I will try to see where the component windpower fits in, how it has been integrated in the system, and its impact on the larger energy system.

Some components of the electric power system are global, others regional, and still others national or local. The global part of the system is *science and electrical engineering* and *standards and certifications* of technical components used in the system. Below these global basic scientific and engineering practices, there are different styles and traditions (Hughes, 1993), often delimited by national borders (Lafferty & Ruud, 2008).

Another global component in the system is the *manufacturing industry*. Some of the largest corporations, such as AEG and Siemens in Germany, General Electric and Westinghouse in the Uniited States, Brown Boveri in Switzerland, and Asea (now ABB) in Sweden, were established already during the very first invention phase of the electric power system. Today they are among the largest transnational corporations in the world.

From a technical point of view, the electric power system is divided into national or regional sub-systems. Such a system is defined as an interconnected electric power system using the same transmission system with a common and identical frequency.

The Nordic electric power system, which covers Sweden, Norway, Denmark, and Finland, is one example. This is in turn a subsystem, which is connected to other systems in Germany, the Netherlands, Poland, Russia, and Estonia. There are interfaces between these subsystems, which allow import and export of electric power.

Connected to these technical systems, there is an organizational structure, with a transmission system operator (TSO) with a legal power to manage the system and order power producers to increase or reduce their production, to keep the system in balance (i.e., to keep production and consumption on the same level). However, in the Nordic power system, there are four TSOs, one for each country, which, however, cooperate closely.

This technical electric power system can be divided into several components:

- *power plants* that produce the electric power;
- *power grid* that distributes electric power to end users; and
- *equipment* that consumes electric power.

There are well-defined interfaces between these components. Directly coupled to these, there are owners/operators of power plants, grid operators, power traders and households/companies that own and operate equipment that runs on electric power.

The technical equipment that is used in power plants and power grids and equipment operated by electric power (motors in industries and trains, household equipment, etc.) and the companies that manufacture such electrical equipment are also part of the electric power system.

There are also legal systems with laws and regulations that govern technical standards, how the system should be operated, and the price of electric power. These systems are in most cases limited by national boundaries. However, some are transnational, such as the Nord Pool regulations, and EU directives on energy.

2.1.5 The electric power system in Sweden

In *När folkhemselen blev internationell* (Högselius & Kaijser, 2007) (in English: *When public electricity turned international),* the Swedish techno-historians Per Högselius and Arne Kaijser start out where Hughes starts, with a description of the development of the Swedish electric power system. Their focus is, however, on how this system has changed after Hughes's book ends (in the 1930s) and how Swedish actors have expanded their business to other countries in Europe.

Their theoretical base is Hughes's so-called large technical systems (LTS) approach. Most of the researchers have applied this approach to describe systems from invention to technology transfer. Högselius and Kaijser take this analysis further, to the stage of *saturation* and *stagnation*. They have also moved from national socio-technical systems to the international level; the internationalization of ownership, international power trade and international regulations.

It describes how the Swedish state-owned power company Vattenfall changed its business strategy and aims from *technical* to *financial* engineering. This was made as the political-economic framework was reformed from a regulated monopoly to an unbundling of the power sector and introduction of free competition and a market place for electric power—Nord Pool (how the Nord Pool market works is described in Chapter 3, Section 3.4, The Electric Power Market).

The critical problems that triggered this change were first the stagnation of electric power consumption in Sweden, and second, the new paradigm of market liberalization, instigated by Sweden's application for membership in the EU.

Two other critical problems also had a great impact on the Swedish power system during the last fifty-year period. The power industry considered the power system to be a purely technical system, and their duty and aim were to operate and develop it in a cost-efficient way. This was actually also done to perfection, although the ownership was mixed: private companies, municipal utilities, and the state-owned authority Vattenfall. The larger power companies had created and institutionalized a well-operating cooperation (Högselius & Kaijser, 2007).

By the end of the 1970s, the power system became a *political* issue, with the introduction and the conflicts about nuclear power. This was to stay on top of the political agenda in Sweden for decades. The power industry was taken by complete surprise by this development.

The next critical problem was the *environmental demands* that were raised, due to the emissions of carbon dioxide, and the climate change issue. Finally, within the power industry, there was a shift from technology to economics, when the power market was deregulated and the Nord Pool market started to operate.

2.1.6 Electric power in the global energy system

A socio-technical system can, as I understand it, be described as consisting of three interacting parts: the technical system (a system of artifacts), a political system (laws, rules, etc.), and an economical system (costs, fees, tariffs), which are embedded in a common organizational structure (companies, public agencies) and a culture (tradition, values, practice). See Figure 2.1.

There are, of course, also other inputs to these three parts from outside the system, from citizens and interest groups to the political system, from new technological innovations to the technical system, and from world market prices and so forth in the economic system.

However, the electric power system is also part of a larger energy system (see Figure 2.2). There is a continuous input of energy from the sun and a

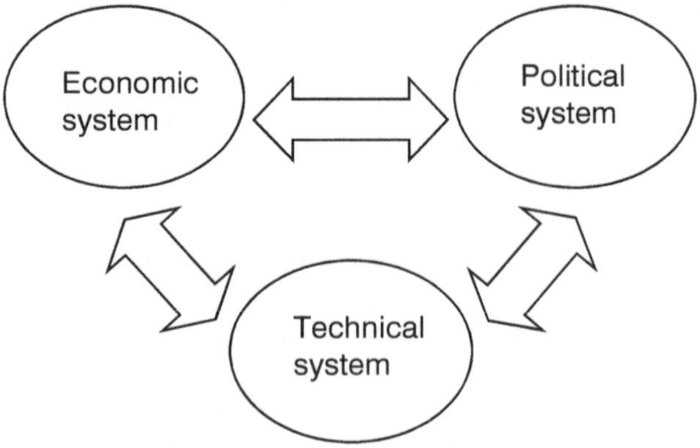

Figure 2.1 Socio-technical system. A socio-technical system can be described as a system with three interacting parts

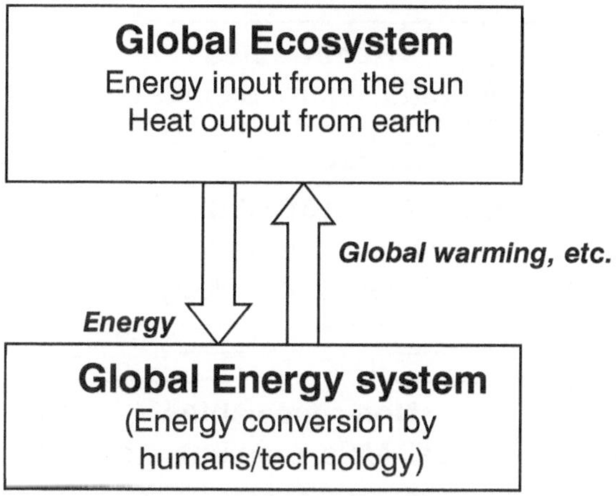

Figure 2.2 Global energy system. The electric power system is part of the global energy system, which interacts with the global ecosystem

similar dissipation of energy (heat) from earth back to space. This input/output should be in balance. However, the global energy system has affected this balance, by emissions of greenhouse gases (GHGs) that cause global warming.

In the LTS model, the relation to the global ecosystem, or the biosphere, is neglected. Factors in the environment are not components of the system, according to Hughes (Hughes, 1989). However, Hughes's concept of *environment* refers to the socio-political and other technical systems in the technosphere and not to the ecological system.

There are two basically different kinds of energy resources that are used in the global energy system (see Figure 2.3).

Renewable energy sources are utilized in real time (sun, wind) or during annual seasons (hydro, some biomass) or the lifetime of biological species (forest biomass). Non-renewables are fossilized biomass from prehistoric eras or radioactive materials such as uranium. The use of non-renewable energy sources changes the balance of the global ecosystem.

The concepts *renewable energy* and *sustainable energy* are often considered as synonymous. These concepts are, however, not identical (depending on who defines the terms), Professor Henrik Lund from Aalborg University points out. Sustainable energy can be defined as energy sources that cannot be expected to be depleted in the time frame relevant to the human race. However, nuclear as well as carbon-capture technology are sometimes included in the definition of sustainable energy but not in renewable energy (Lund, 2010).

These facts and cause–effect relations between the global energy system and the global ecosystem are not social constructs but governed by the laws of nature, which are non-negotiable.

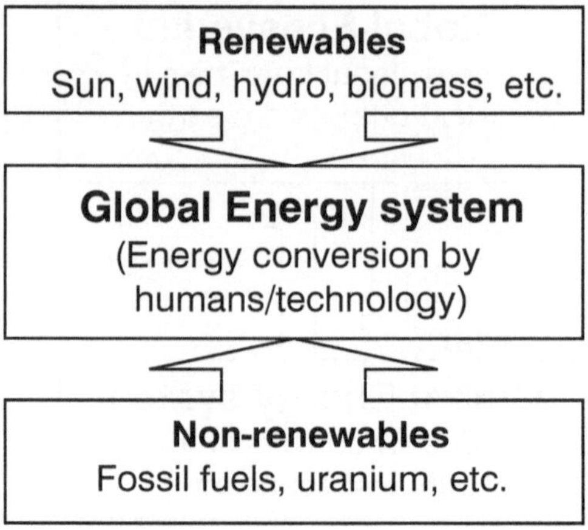

Figure 2.3 Renewable vs. non-renewable energy sources. The energy sources used in the global energy system and the electric power system can be either renewable or non-renewable. Renewables do not have any impact on the global ecosystem in the form of global warming

2.1.7 Overlapping systems

The electric power system, the transport system, and the heating system are interconnected and overlapping subsystems in the global energy system, with common components (raw materials, etc.), technology (motors, boilers, etc.), and environment (the global energy system, etc.). Another large technical system that has had a direct impact on the electrical power system is information technology (IT). Without integration of components from IT, a deregulation of the power system would have been impossible, as a fast transfer of very large amounts of information is a precondition for a market with independent actors.

Other important aspects and concepts for a discussion of the global energy system are scale, availability, transport, and trade. The energy system varies from local use of local resources (firewood) to use of resources refined in factories, traded on the world market, transported around the globe, and used in very large systems (thermal power plants fired by oil/coal, nuclear reactors using uranium) that distribute energy over large areas. Some parts of the global energy system are in fact not parts of any large technical systems at all. In the countryside in many developing countries, firewood is the only energy source that is available.

Models, such as the LTS-model, are tools used for analysis and are of use if they give a proper understanding of the reality the model describes and explains. It is neither the aim nor the topic of this thesis to develop any new models, but I think that the global ecological system, the biosphere, should be incorporated in LTS-models. Today there is an obvious and strong interaction between most large technical systems and the biosphere, and these interactions can also be quantified and given economic values. It would also make it easier to evaluate different development paths and policies for such systems.

2.1.8 Social-ecological systems

With the concept of social-ecological systems, natural resources and ecological systems are integrated with social actors and institutions. Social-ecological systems are complex and adaptive and delimited by spatial or functional boundaries surrounding particular ecosystems and their problem context (Glaser, Krause, Ratter, & Welp, 2008).

According to Redman et al. (Redman, Grove, & Kuby, 2004) a socio-ecological system can be defined as

- a coherent system of biophysical and social factors that regularly interact in a resilient, sustained manner;
- a system that is defined at several spatial, temporal, and organizational scales, which may be hierarchically linked;
- a set of critical resources (natural, socioeconomic, and cultural) whose flow and use is regulated by a combination of ecological and social systems; and
- a perpetually dynamic, complex system with continuous adaptation.

The concept of socio-ecological systems emphasize the integrated concept of humans in nature and stress that the relation between social systems and ecological systems is artificial and arbitrary (Berkes, 2001).

Elinor Ostrom, who was awarded the Nobel Prize in 2009, has also used and discussed this concept (Ostrom, 2007). Her main focus has been research on common-pool resources (CPRs) characterized by renewable resources (forests, water irrigation systems, fisheries, grazing lands), scarcity, and where participants are dependent on one another (Ostrom, 1990).

Ostrom shows that stable institutions of self-government exist and can manage scarce resources to avoid overconsumption. These institutions, if certain conditions are fulfilled, are often more efficient for sustainable management than the proposed "cure-all" solutions proposed by some other scholars, such as privatization of property rights, or enforcements imposed by outside force, such as government ownership, regulations or monitoring. Her research also refutes that the *tragedy of the commons*[2] (Hardin, 1968) is unavoidable (Ostrom, 1990, 2005, 2007).

Some windpower ownership models are institutions for common use of the natural resource wind energy. However, the wind cannot be considered a scarce resource and thus falls outside this strict definition of small-scale CPR utilization.

Ostrom and her colleagues have, however, also developed a more general model, or rather conceptual framework, called the institutional analysis and development (IAD) framework, to analyze complex social-ecological systems (Ostrom, 2005).

The purpose of this framework is to

> ...help integrate work by political scientists, economists, anthropologists, lawyers, sociologists, psychologists, and others interested in how institutions affect the incentives confronting individuals and their resultant behavior.
>
> (Ostrom, *Understanding Institutional Diversity*, 2005, pp. 8–9)

The components of this framework, described by Ostrom as a multitier conceptual map, are *exogenous variables*, an *action arena*, and *outcomes*.

In a seminar 'Science and Technology for Sustainable Well-Being' in 2007, Ostrom presented a paper on social-ecological systems (SESs) (Ostrom, 2007). Ostrom presents a development of the IAD framework to encompass social ecological systems in general:

At this broad level, one can begin to organize an analysis of how attributes of

- a resource system (e.g., fishery, lake, grazing area);
- the resource units produced by that system (e.g., fish, water, fodder);
- the users of that system; and
- the governance system.

jointly affect (and are indirectly affected by feedback from) the patterns of interactions and resulting outcomes achieved at a particular time and place, and how these may affect and be affected by larger and smaller socioeconomic and political settings in which they are embedded as well as by larger and smaller SESs.

Ostrom hopes that further use of this framework will make it possible for researchers to answer three broad questions:

1 What patterns of interactions and outcomes are likely to result from a particular set of rules for the governance, ownership, and use of a resource system and resource units in a specific technological, socioeconomic, and political environment?
2 What is the likely development of different governance arrangements, use patterns, and outcomes with or without external financial inducements or imposed rules?
3 How robust and sustainable is a particular configuration of users, resource system, resource units, and governance system to external and internal disturbances?

This framework for analysis of SESs is shown in Figure 2.4.

More than thirty different variables that have impacts on interactions and outcomes have been identified.

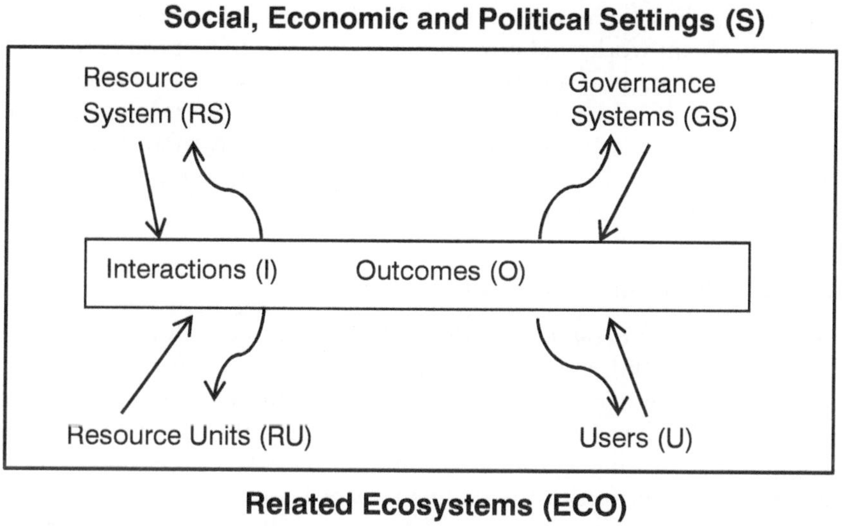

Figure 2.4 Framework for social-ecological systems. Multitier framework for analyzing social-ecological systems. Straight arrows represent direct casual links, curved arrows represent feedbacks

All of these variables are, of course, not relevant for any particular research question. The wind is a renewable resource and most likely easier to analyze than many other SESs. Ostrom stresses that it is the *combination* of variables that is important. The spatial and temporal aspects should also be included.

I find it interesting to see how this framework can be used to analyze the research questions posed in this thesis, and whether it is possible to recognize a combination of variables that lead to a sustainable and productive use of the wind resources.

Ostrom and her colleagues working with CPRs have found some variables that are important for a sustainable management: face-to-face interactions and attachment to place, which may be relevant in this context.

To get the point of renewables versus non-renewables, the relation and interaction between several SESs will be necessary. This may, however, be too big a question to manage within the framework of this thesis.

This framework and its variables are described very shortly. A detailed description is presented in Poteete et al., *Working Together* (Poteete, Janssen, & Ostrom, 2010).

2.2 Transition to renewable energy

There is a global consensus, expressed in international treaties such as the Kyoto protocol and EU targets, that a transition to renewable energy is both urgent and necessary to stop or at least delay and limit a global climate change. How this should be done, and which strategies should be used, which targets aimed at, and how this "burden" should be shared by different countries are, however, controversial. Many scientists have analyzed how the energy system works and how such a transition can be promoted.

In this context, it is interesting to investigate whether ownership of different kinds of power plants, conventional as well as renewable, could have an impact on a transition to a renewable energy system. Here is a short overview of some of this literature on transition, concepts used in this context, and an interpretation of its relevance.

2.2.1 *Path-dependence and lock-in*

In *An Evolutionary-Economic Analysis of Energy Transitions* (van den Bergh & Oosterhuis, 2008), van den Bergh and Oosterhuis define the concepts *path-dependence* and *lock-in* in the context of evolutionary economic theory.

Path-dependence is created by repeated selection of a specific technology that results in increasing returns, based on scale-advantages, learning-by-using, imitation, and other factors. Increasing returns lead to larger market shares, and this growth will be the cost of others. It is a self-reinforcing mechanism, which can give dominance to a particular technological or economic regime. This is known as *lock-in*. Lock-in is regarded as a fundamental barrier to intentional transitions, like a transition to a renewable energy system.

In *Barriers and Options for Future Energy Transitions: Lessons from a Historical Analysis of the Dutch Electric System* (Verbong & Geels, 2008), the Dutch researchers Geert Verbong and Frank Geels give an interesting view of developments in Holland, parallel to the description of the Swedish power system in *När folkhemselen blev internationell* (Högselius & Kaijser, 2007)

Verbong and Geels have identified several mechanisms that contribute to *stability* and *lock-in*:

- cognitive routines that blind engineers to developments outside their focus;
- regulations and standards;
- life style adaption to technical systems;
- sunk investments in machines, infrastructure, and skills;
- organizational commitments;
- vested interests;
- social resistance to new technologies;
- mutual role expectations; and
- organizational capital in networks.

These factors create *path-dependencies* that have deep historical roots. Verbong and Geels then apply a multilevel perspective on the transition of the electric power system in the Netherlands. This perspective distinguishes between three levels: micro, meso, and macro, and they focus on the meso-level. They call this the *socio-technical regime*, the level that accounts for the stability of the system. This regime consists of three interlinked elements:

- a network of actors and social groups;
- formal, cognitive, and normative rules; and
- the technical elements.

The concept *socio-technical regime* is very similar to the concept *technological style*, which is used by Hughes in *Networks of Power* (Hughes, 1993) (see Section 2.1.2, this volume).

The micro-level is formed by *technological niches*, which are created by small networks of actors who are willing to invest time and money in the development of novelties. The macro-level is the *socio-technical landscape*, which influences developments in niches and regimes. Transitions come about through interactions between all three levels.

According to historical studies made by the authors and other researchers, radical technological change follows a specific pattern: Radical innovations first emerge in a small niche, where they can develop and stabilize. If powerful social groups support the innovation, and if its price/performance ratio is improved, it may enter mainstream markets. This wider diffusion does not depend only on niche-internal drivers, but also on wider ongoing processes at the regime and landscape levels. Landscape changes can create pressures on the regime, resulting in problems and tensions:

a) There may be technical problems or decreasing returns;
b) there may be changes in cultural values and concerns about certain side effects of existing technologies;
c) there may be changes in markets and user preferences; and
d) policy makers may introduce stricter regulations that cannot be met by the existing technology.

Such regime processes may create "windows of opportunity" for niche-innovations to break through more widely (Verbong & Geels, 2008, p. 179).

The development of windpower from marginal small wind turbines for farmers to the fastest growing technology for electric power production on the global market is a good example of a technology developed from a small niche into mainstream technology, a development described in greater detail in Chapter 4, Windpower Development and Ownership.

Actually, all factors from a) to d) listed above have contributed to changes in the socio-technical landscape. A recent example is the decision by the German government to close old nuclear power plants and decommission the rest by 2022 (TT-AFP, 2011) as a result of the nuclear power catastrophe in Fukushima in Japan in 2011, due to b) ... concerns about certain side effects of existing technologies.

2.2.2 *National socio-technical regimes or cultures*

The fact that this nuclear catastrophe hardly had any impact at all on energy policy in Sweden is one example that indicates that socio-technical cultures have national borders. The same conclusion can be drawn from a description and comparison of the position of windpower in the *electric power cultures* in the neighboring countries of Sweden and Denmark.

In Denmark, the first wind turbines for electric power production were developed already in 1892 by Paul laCour (Nissen, 2003) and in 1957 a new effort was made with the first grid-connected wind turbine, a 200 kW prototype in Gedser, which operated until 1967 (Nelson, 2009). Danish engineers have thus, by tradition, been aware of windpower, and some members of the electric power community had good knowledge and experience of this technology. Danish technical universities have since then also had researchers and education programs on windpower, which made it possible for Denmark to play the role of pioneer for the development of this renewable energy source.

In Sweden, the scientific and engineering communities, the Royal Academy of Science (KVA) and the Academy of Engineering Science (IVA), have a strong commitment to nuclear power.[3] Windpower has, with a few exceptions, not become a regular part of the curriculum in the education of electric power engineers. Only during the last few years have some courses been added at Uppsala University and at the Royal Technical University (KTH) in Stockholm.

Outside of this old traditional power engineering community, the new university in Halmstad developed courses on windpower technology, and the small provincial university on Gotland developed courses for windpower project developers. In 2010, a windpower development center was established at Chalmers Technical University in Gothenburg, initiated by the Swedish Energy Agency (i.e., the state), so there may be a trend toward a new more positive attitude toward windpower in these communities of practitioners. This did, however, happen first when windpower had become a mainstream technology on the international electric power market.

2.2.3 *Grassroots initiatives and private ownership*

In *Regional Diversity of Wind Energy Initiatives: The Dutch Experience* (Bruinsma, 2008), two Dutch researchers, Frank Bruinsma and Ron Vreeker, discuss the development of wind energy in the Netherlands. Windpower is the most important renewable energy source used in the Netherlands, and in October 2006 (when the book was published), there were more than 1,800 wind turbines with a nominal power of 1450 MW on line, which produced some 2.7 percent of the electric power used in the Netherlands. This was accomplished by investment subsidies and fiscal advantages.

In 2001, the government agreed with the twelve provinces in the Netherlands to install 1,500 MW by 2010, with the target divided between each of the provinces. By 2006, the target for 2010 was almost reached, and it was surpassed with a good margin in 2010; however, some provinces performed far below the targets that had been set, while others surpassed them with a large margin. The province of Utrecht, with a target of 50 MW, had realized only 0.2 MW, while Flevoland, with a target of 220 MW, had developed 587 MW in 2006 (Bruinsma, 2008).

According to Bruinsma and Vreeker's analysis, the different success of Utrecht and Flevoland can be explained by the different planning procedures. In Utrecht, a top-down approach was used, while Flevoland applied a bottom-up model. In Flevoland, projects were initiated by farmers, and the regional authorities facilitated the process.

This implies that farmers were owners of the wind turbines and that ownership is a crucial factor for development of windpower and for other renewable energy sources as well. Local initiatives by private citizens, in this case farmers, can promote a transition to renewables.

2.2.4 Promotion of renewable energy systems

In the book *Promoting Sustainable Electricity in Europe* (Lafferty & Ruud, 2008), edited by William Lafferty and Audun Ruud, promotion systems in some different EU-member states and Norway are described and evaluated. The overall aim is to evaluate the Renewable Energy System-Electricity (RES-E) directive from the EU, and the "indicative targets" set for the individual member states.

The EU Commission was focused on the so-called techno-market approach[4] to implement development of renewables, which according to Lafferty and Ruud was neglecting contextual variables within the individual countries. Their analysis, and the case studies from the Netherlands, Denmark, Ireland, Spain, Finland, Austria, Sweden, and Norway, led to the new concept *path creation*, as opposed to *path dependence*.

The outstanding model for path creation is Denmark and the success story of windpower development and the growth of the windpower industry there (Karnoe & Buchhorn, 2008). This path creation can be described as a virtuous circle, going from technology development to industry development and market deployment by interaction of a technology cycle (technology progresses through innovation, feedback, and R&D investments) cost reductions, efficiency increases, enhanced applicability, and a market cycle (cheaper, better, and new products enlarge markets) greater sales increase demand and greater use gives more feedback. These cycles interact so that stimulation of the production results in higher quantity as well as quality.

This mode of operation has worked in Denmark and a decade later also in Germany, but why does it not work in all EU member states? This is, according to Lafferty and Ruud, due to the character of the dominant energy system (DES). The DES is viewed as a constellation of energy-related actors—investors, resource exploiters, market facilitators, energy producers, energy system managers, public regulators, and so on—who influence the potential for change.

The actors of the DES thus define what Verbong and Geels call the socio-technical regime (see Section 2.2.1, this volume).

On one side, there is a path-dependent regime of the DES (nuclear, hydro, fossil fuels, etc.); on the other, the challenging regimes of new technologies (wind, biomass, etc.). The new regimes must get the prices right to be competitive, but what is right and how the techno-market factors work varies in different countries, due to historical, cultural, social, and political settings. How the indicative targets set by the EU are implemented, thus depends on the *national contexts*.

This book was published in 2008 and describes the situation in the years before then. Since then, windpower as well as biomass power plants have taken the step from niche to mainstream technology (see Chapter 4, Windpower Development and Ownership) and are rapidly integrated into the DES, at least in many countries.

The fact that, for example, the Swedish state-owned power company Vattenfall makes large investments in windpower plants in the United Kingdom, the Netherlands, and other countries, but more marginal windpower investments in Sweden, also stresses the relevance of national contexts. After Vattenfall started to operate as an international company, it chose to invest in windpower in countries where the support systems are most favorable, with fixed high prices.

It is also interesting to note that the ownership structures and models also differ significantly in different countries, which will be described in more detail in Chapter 4.

2.2.5 Preconditions for radical technological change

In the book *Renewable Energy Systems* (Lund, 2010), Henrik Lund from Aalborg University in Denmark offers a number of useful definitions as well as a theory on *choice awareness*. He also has an interesting discussion about the market and on economic benefits for society from radical technological change options.

Technology is defined as one of the means by which humankind reproduces and expands its living conditions. It embraces four elements: *technique, knowledge, organization*, and *products*. *Radical technological change* is defined as a change of more than one of the four elements of technology.

Lund makes a distinction between *concrete institutional economics* and *applied neo-classical economics*. The second concept is defined as neo-classically based methods, such as cost-benefit analyses and equilibrium models, applied to existing real-life market economies. This is opposed to theoretical models of an assumed free market.

Institutional economics focuses on the understanding of the role of human-made institutions in the shaping of economic behavior. *Concrete institutional economics* deals with the concrete institutional conditions that form the development of a specific society.

According to Henrik Lund, there are at least three political arguments to promote and develop renewable energy sources:

- *Energy security*: The main reason is to get rid of oil dependency. For Sweden also, nuclear fuel dependency is relevant.
- *Economics*: Job creation, industrial innovation, and balance of payment. This was the main political driving force behind the industrial development of windpower in Denmark in the 1980s and 1990s, according to Lund.
- *Environment and development*: Climate change became a key issue in Denmark (and the rest of the world) in the 1990s, and its importance has increased since then.

To promote and implement radical technological change in the energy system, the whole system has to be considered (i.e., available technology as well as institutional frameworks and public regulations).

Lund presents a number of case studies, most from Denmark but also from Thailand and the United States.

> From these cases, it can be observed that organizations linked to existing technologies will initiate project proposals within their organizational frameworks. One cannot expect alternatives representing radical technological change to originate from such organizations. It is outside their discourse, and it is not within their interest or perception. ....., even when political decisions have been made implying the wish of radical technological change, the organizations linked to existing technologies will continue to initiate project proposals within their organizational framework.
>
> (Lund, 2010, p. 241)

And further:

> For society, it is not an efficient public regulation measure to force existing organizations to act at variance with their interests. Such a measure has to be supplemented by institutional changes.

Lund's conclusion is that radical technical change, for example to develop a renewable energy system, has to be initiated by actors *outside* the power industry. Another important point is that different systems have to be analyzed, not only from a strict technical and business economy point of view. The institutional frameworks, such as taxes, tariffs, and the like, as well as political goals, such as high employment and balance of payment, have to be included. Alternatives, which are better from an environmental and also a socioeconomic point of view than those proposed by the power industry, can often be found, if the analyses are made in this more comprehensive way.

> In all cases it was possible to identify alternatives based on radical technological changes that were likely to fulfill the political objectives better than the proposals of existing organizations. However, such alternatives were presented by others, such as universities, NGOs or local citizens.
>
> (Lund 2010, p. 249)

The development of windpower is often described as a success story for Denmark. An interesting point is that the Danish power companies and grid operators were strongly opposed to windpower, at least up to the year 2000 (SEAS, 2002).

In Denmark (as in Germany), the politicians were stronger than the power companies and forced them to accept an agenda they actually opposed. In both these countries, windpower was developed, and investments made, by local citizens, cooperatives, and companies outside the power industry.

Another interesting case in this context is Spain, where power companies and industry have been very active actors and have contributed to the windpower success story of Spain during the last decade, which shows that the conclusion of Henrik Lund above not necessarily applies to all actors in the DES. The radical technical change was, however, initiated by an actor outside the power industry, by the state agency National Institute for Energy Saving and Diversification (IDEA). And the success of windpower in Spain can be explained by the ownership model public–private partnership (PPP) (Dinica, 2008). A more detailed description of the development in Spain is given in Chapter 4.

2.2.6 *Energy policy*

The preconditions for development of windpower are set by the national policies in a country and by the laws and regulations that are inaugurated by the parliament. The politicians also decide the economic rules of the game. Taxes, charges, subsidies, and other means of control often form a large part of the power price, on top of the actual cost of power production. There are also different ways to promote windpower and other renewable energy sources. The politicians set the framework for windpower development and in this way will influence the pace of transition to renewable energy systems.

Each country has its own laws that prescribe the procedures that are necessary to get permission to install wind turbines and set the framework for the economic conditions for the sale and distribution of wind-generated electric power.

In Europe, the development of commercial windpower started in the early 1980s. Today, countries such as Denmark, Germany, and Spain have several thousands of MW windpower on line as well as a new industry with thousands of employees. In countries sch as the Netherlands, the United Kingdom, and Sweden, development has been considerably slower and started to take off first around 2005–2006. The explanation for the large differences in the pace of development between different countries can be found in the windpower policy that has been conducted by governments.

With some thirty years of experience, it is quite clear what kind of political measures, rules, and regulations promote a fast development of windpower (and other renewable energy sources) and what measures can be used to keep development on a slow and low level.

In Denmark, Germany, and Spain,[5] it has been profitable to invest in, to own, and to operate wind turbines. The laws and regulations in these countries guarantee a stable and rather high price for the power produced during the time period that it takes for the owners to get their money back. There have been clear political signals that windpower should be developed quickly, and it has not been too difficult and time-consuming to get the permissions necessary to install them and connect them to the power grid (Wizelius, 2007).

The politicians in these countries have had several different motives for their support to windpower. Renewable energy from windpower will reduce the impact from power production on the environment as well as the emissions of carbon dioxide that most countries have been obliged to reduce in accordance with the Kyoto protocol and EU targets. Windpower will enable the politicians to attain these goals.

An equally important factor is probably that windpower by some governments has been considered as a new and very promising industry, with a huge growth potential that can give a considerable payback to the economy by increasing employment, economic growth, and export revenues. In Germany, the support of windpower has also favored their agricultural policy. Wind turbines give an additional income to farmers so that they can make a living and stay in the countryside instead of moving to town and lining up among the unemployed (Wizelius, 2007).

Even if the economic subsidies to windpower during its early stage of development are rather expensive, the politicians have calculated that in the longer run, it will generate benefits to the economy. In countries where the demand for electric power is increasing, such as India and China, there are no other motives necessary to develop windpower, as it is one of the cheapest ways to produce electric power today at sites with good wind conditions. According to an estimate by the Royal Bank of Scotland, windpower on land was the cheapest technology for new power plants in 2010. Scotland has among the best wind resources in Europe, and the economic results depend on the wind resources available.

Such a comprehensive view, which considers industrial policy and national economics, has been lacking in the Netherlands, the United Kingdom, and

Sweden during the decades when the windpower industry was developed. In these countries, the domestic wind turbine manufacturers languished owing to the lack of a growing and stable domestic market. The rules for power sale and the permission processes have been complicated, and the purchase price for wind-generated electric power has been low (Jacobsson & Johnson, 2003).

One of the aims of such a policy has been to reduce the cost of wind-generated power to the lowest possible level, with the legitimate ambition to strengthen the competitiveness of windpower. Instead, the development pace has been retarded, and the domestic manufacturing companies have been nipped in the bud.

2.2.7 Promotion systems for renewable energy[6]

In the project RE-XPANSION, researchers from several European countries have described and evaluated the different support schemes that have been used to promote windpower and other renewable energy sources in European countries. The result of their comparative analysis was presented in 2005 in the report *Support Schemes for Renewable Energy* (EWEA, 2006).

Incentives used to promote windpower can be grouped into three categories:

- *Green marketing*: Voluntary systems where the market determines the price and the quantity of renewable energy.
- *Fixed prices*: Systems where government dictates the electricity prices paid to the producer and lets the market determine the quantity.
- *Quotas*: Systems where the government dictates the quantity of renewable electricity and leaves it to the market to determine the price.

The first of these, green marketing, has proved inefficient; fewer than 1 percent of power consumers actually choose this option.

Systems with fixed prices or quotas are regulated by law and are thus mandatory, which makes them more efficient. The report defines five different types of support schemes for renewable energy:

- investment subsidies;
- fixed feed-in tariffs;
- fixed premium systems;
- tendering systems; and
- tradable green certificate systems.

Investment subsidies have been used by many countries in the early stages of windpower development, for example Denmark, Germany, and Sweden, and have proved to be quite efficient. In the United States, an investment subsidy was introduced in 2009 when the U.S. Congress passed *The American*

Recovery and Reinvestment Act to address the growing shortage of finance for renewable energy projects in the wake of the financial crisis (Bolinger, Wiser, & Darghouth, 2010).

The advantages are that it is a simple system and that the subsidies are paid up front. The support cannot be reduced or withdrawn during the lifetime of the project and thus gives security to the investor. The drawbacks are that it does not differentiate good projects from bad. Investment subsidies have now been abandoned in most countries.

Fixed feed-in tariffs have so far been the most efficient way to promote windpower. The price paid for wind-generated electric power is fixed, either to a specific value or in relation to the consumer price. The price can be fixed during the lifetime of the turbine or until a specified target (i.e., a number of full-load hours[7]) is reached. The system always guarantees that the investors will get their money back. It is used by the three countries that have been very successful when it comes to windpower development: Denmark, Germany, and Spain.

Fixed premium systems have been used in combination with other promotion strategies. The premiums, in Sweden called *environment bonus*, give a premium for the avoided costs on health and environment. In practice, the value of the premium is set in relation to the power price to make windpower competitive.

Tendering systems have been used in the United Kingdom, in the so-called non-fossil fuel obligation system. It did not work out too well, as many of the projects that won the tenders never were built. It has now been replaced by a green certificate system. Tendering is, however, used for offshore developments in Denmark, the United Kingdom, and the Netherlands.

Tradable green certificate systems for renewable energy have been introduced during the last years in Italy (2002), the United Kingdom (2002), Belgium (2002), and Sweden (2003). Producers will get certificates in relation to the power produced during a year, and these certificates can then be traded on a certificate market where the price is set by supply and demand.

In the EWEA report, these different support schemes have been evaluated through a survey among more than 500 experts from the energy field. According to this survey, the most important properties of support systems are *investor confidence* and *effectiveness*.

In this survey, the feed-in tariff system got the top score, with green certificates and tendering at the bottom. The green certificate system got low scores for the most important criteria—investor confidence, and effectiveness—as compared to other support schemes.

The rules and regulations for these different systems vary, so each country actually has its unique system. This makes it difficult to compare and evaluate the efficiency of the different systems.

From a historical point of view, it seems obvious that the fixed price systems are most efficient, as they have made development in Denmark,

Germany, and Spain so successful. However, there are other countries that have used the same system with less success, such as Greece.

The support system is but one of several factors that has to be right. According to the authors of the report, there are four main ingredients in a potentially effective overall promotion strategy for renewables:

- well-designed payment mechanisms;
- grid access and strategic development of grids;
- appropriate administrative procedures and streamlined application processes; and
- public acceptance.

The systems with tradable green certificates have now been in use for some years. The mechanisms that did not work so well have been improved. In Sweden, the time frame and the quotas have been adjusted several times since the system was introduced.

There have been quite strong arguments for a change from the fixed price system to the tradable green certificate system after the deregulation of the market. The quota system is claimed to be more market-oriented than the fixed price system. The authors of this report have a different opinion:

> A system where the government fixes quantity and leaves it to the market to determine the price is unlikely to be more "market oriented" than a system where the government fixes the price and leaves it to the market to determine the quantity.
>
> The main difference between quota based systems and price based systems is that the former introduces competition between the electricity producers (e.g. wind turbine operators). Competition between manufacturers of plant (e.g. wind turbines), which is crucial in order to bring down production costs, is present if government dictates either prices or quantities.
>
> (EWEA, 2006)

It is possible to achieve a specific target (MW new power) to a specific time by giving a guaranteed price for new power produced that will stimulate the necessary investments. This is exactly what Denmark has done with windpower. In Denmark, development of windpower actually has reached the goals that have been set several years ahead of schedule. The target set for 2005 was reached already in 1999 (Wizelius, 2007).

2.2.8 *Support schemes in non-European countries*

The countries with the fastest growth and most installed windpower capacity are China and the United States. India has been one of the pioneers in windpower development, and there are many other non-European countries with an impressive development of windpower, such as Canada, Brazil, Mexico, Egypt, and Morocco to name a few. Each of them has its own support scheme, but they can be categorized the same way as those used in the European Union, and are described in Section 2.2.7.

India

India was the first country in Asia to develop windpower on a commercial scale. By the end of 2012, India had installed more than 18,000 MW and is still one of the countries in the world with most windpower on line. Only China, the United States, Germany, and Spain have more. India also has a strong domestic wind turbine manufacturing industry, which has more than two decades of experience in this field.

Windpower has become commercially viable in India, and 97 percent of the investments in windpower are made by the private sector, which indicates that windpower can successfully compete with other sources of energy.

In the eighth five-year plan, running from 1990 to 1995, the government of India set a target of achieving 500 MW through private sector participation. To achieve this target, the government in 1990 introduced a package of incentives to create an attractive investment climate for windpower. The most important incentive was an accelerated depreciation of 100 percent in the first year for investments in windpower.

In Tamil Nadu, in the south of India, where most windpower is in operation, the Electricity Board introduced a policy on *banking and wheeling*[8] of electricity from windpower plants. This enabled so-called *captive consumption*: companies could produce their own power and transmit it through the grid to their factories.

In 2003, a new Electricity Act was enacted to reform the electric power sector. Each state has a regulator, the state Electricity Regulatory Commission. A regimen of interstate trading, open access, and availability-based tariffs was initiated, and the commissions have set tariffs for purchase of electricity from windpower plants. The commissions also have to determine a quota for the renewable purchase obligation. The policy obligates each retail seller to include in its resource portfolio a certain proportion—quota—of power from renewable energy resources such as wind, solar, small hydro, and biomass energy (Earnest & Wizelius, 2011).

In 2008, a generation-based incentive was introduced. This could be utilized only by those who did not use the accelerated depreciation benefit. In March 2011, a renewable energy certificate system was introduced as well. This was introduced to enable states to fulfill their renewable purchase obligations, and it has become one of the key growth drivers for renewable energy in India (Bhambri, 2011).

China

In 2003, the electric power market in China was deregulated. The huge state-owned State Power Corporation was split into two grid-operating companies, five power generation companies, two consulting companies, and finally two companies for construction of hydro power plants. These are still huge state-owned companies, but they act independently. Power production and grid operation (i.e., transmission of power) were separated. The grids are controlled by the government, but power production is open also for independent power producers and private investors.

In 2004, China announced the ambitious goal to generate 16 percent of its energy from renewable energy sources by 2020. The government introduced targets, mandatory market share requirements, and economic incentives for renewables and created a billion dollar market, in particular for windpower, which is the most technically mature and cost-efficient renewable energy technology. Since then, there has been a boom in the renewable energy sector in China.

In the 1990s, windpower projects were funded by overseas donors. The tariffs were very low and hardly covered the operation and maintenance costs. From the late 1990s to 2002, local governments set the tariffs, which had to be approved by central government, with widely varying prices in different regions. From 2003, the prices for the so-called national concession projects, which have to be larger than 50 MW, were decided by competitive bidding, and the National Development and Reform Commission (NDRC) is in charge of approval. Besides tender-based pricing, there can be an approved price set by the government. Projects of less than 50 MW are under the responsibility of the provincial governments.

The concession program was introduced in 2002. The NDRC selects sites and invites a bidding process where the lowest bid gets the right to develop the site. The government then guarantees a fixed price for the first 30,000 hours of operation. After that, local electrical prices are adopted. Several rounds of bidding for concessions have been held so far, and it has been quite successful.

However, the system has had its drawbacks, as bidders have engaged in a fierce price competition, so that many wind farms do not make ends meet. The concession program is dominated by the large state-owned companies, which can subsidize windpower plants with revenues from other more profitable power plants. Investments are made to fulfill mandatory market shares and other non-economic motives. Another fact is that only 28 percent of the concessions awarded in 2004–2006 have actually been installed. This is quite similar to what happened in the United Kingdom with the tendering processes used in the 1990s.

The criteria for awarding concessions have been continuously revised since 2005 to avoid these problems. Owing to the high cost of participating in the bidding for national concessions, many project developers opt for provincial projects, with less than 50 MW capacity. The tariffs for these projects are set by the provincial government but have to be approved by the pricing bureau of the NDRC.

The renewable energy law came into effect on the first of January 2006. It serves as a framework law with the aim of increasing the share of renewables in China's energy mix, promote domestic industry, and support regional economic development. The details will be set by regulations for implementation, which makes it easy to change the rules when called for.

Grid operators are obliged to connect all renewable power plants and to buy the power as well. The feed-in tariffs are decided by the NDRC price committee and have to guarantee economic viability.

The medium- and long-term plans for renewable energy development were published by the NDRC in September 2007. Already in February 2008, the targets were revised upward. In this plan, the share of renewable energy in the power mix has been set, and targets for windpower development have been quantified, specified as installed capacity. The plan also sets quotas for the share of renewables in the energy mix of power companies: one quota for small producers and a higher one for power companies, with a total capacity of more than 5 GW.

The National Renewable Energy law was revised in 2009, adding that the grid operators had to purchase guaranteed quantities of electricity generated by renewable energy. Fixed feed-in prices were also guaranteed for the lifetime of projects.

In August 2009, China replaced its centrally controlled bidding system with a new series of feed-in tariffs for all onshore wind farms built after this date (Wizelius & Earnest, 2009; Pengfei, 2011).

United States

In the United States, the federal government has supported windpower development through a mix of research and development funding, tax subsidies, direct grants, and low-cost financing.

In the late 1990s, the U.S. Congress introduced the Production Tax Credit, which provided performance-based incentives. This helped to drive a rapid growth of windpower development, although it has been enacted only for a few years at a time and contributed to a boom-and-bust cycle. Tax credits can be utilized only by companies with a balance sheet.

Some states (29 states, District of Columbia, and Puerto Rico, in February 2011) also have renewable portfolio standards that require utilities to purchase an increasing fraction of their power from renewables. This system is similar to the green certificate/quota system used in Sweden. It has been a driver of windpower development (Veilleux & Hamrick, 2011).

2.2.9 *Non-technical barriers*

Even if the support systems are well designed and the economic conditions make investment in windpower safe and viable, there may be difficulties to develop new windpower projects. There may be so-called non-technical barriers that prevent or delay the development.

According to the European Wind Energy Association:

> The barriers faced by developers during this process (project development) are often related to the approval and scope of the Environmental Impact Assessment, compliance with spatial planning, the number of parties/authorities involved and to barriers related to other stakeholders involved in the process (e.g. social acceptance issues).
>
> The barriers faced by developers during this process (grid connection) are mostly related to an absence of clear information on the available grid connection capacity, a lack of planning for future grid extension and reinforcements on behalf of system operators, insufficient grid capacity, and other aspects such as land ownership and the EIA.
>
> (EWEA, 2010)

Miguel Mendonca from the World Future Council presents a more extensive list of barriers to development of renewable energy systems in general:

- large public subsidies for fossil and nuclear fuels, and lack of assessment of their future costs;
- opposition from utility companies that are intolerant of competition from other investors;
- government preference for large centralized solutions;
- lack of administrative experience with renewables; and
- investment risk and uncertainty.

These and other factors combine to distort the conditions for market access for renewables, he argues (Mendonca, 2007).

Such non-technical barriers have been and still are a problem in many countries, and also in Sweden.

In Sweden, the government has made several attempts to remove these barriers but without any success. The reforms implemented in 2007 (The Swedish Government, 2006) to make the permission process smoother and faster had the opposite effect. The time from the submission of an application until permission is granted has increased. One reason for this is that the municipalities have got a right to stop any windpower project by a veto. Another is the very slow processing of applications at the county administrations and environment courts as well as the demands on research to be made by developers to prove that impacts on nature and protected species are acceptable.

In 1990, the Swedish geographer Inga Carlman published a thesis that showed that demands on windpower projects were much more stringent than for other power plants that need permission and that the positive impacts from reduced emission of GHGs and other gases that cause damage to the environment are not accounted for (Carlman, 1990).

In a thesis from 2004, *Wind Power, Landscape and Meaning—A Study on Wind Power and Human Spatial Preferences*, Tom Böhler concludes that the debate about whether windpower should be used in a landscape or not generally concerns aesthetics and that important aspects such as the connection between local renewable energy production and global environmental responsibility seldom are taken up (Böhler, 2004).

This applies to the judgments in the permission processes as well, just as Inga Carlman found in the 1980s.

2.2.10 Transition in Sweden

The DES in Sweden with hydropower and nuclear power has been supported by comprehensive governmental programs promoting their expansion as part of the general industrial and energy policy. Actors in the DES—manufacturing companies, utilities, research institutions, regulatory bodies (i.e., the culture in Hughes's terms)—with vested interests in the DES have overlooked or perceived new energy technology as a threat (Kaijser, 1988).

This explains why political energy agreements in parliament about a promotion of electricity generation from new renewable energy sources such as wind, solar energy, small-scale hydro, and biomass (The Swedish Government, 1991, 1997) during thirty years from the early 1990s did not materialize.

A large-scale development of windpower in Sweden started first around 2007. By then, several neighboring countries such as Denmark and Germany had a large share of windpower in their power systems; and by technical development, the wind turbines had grown, which made it possible to develop windpower plants with large capacity. The support system with quotas and certificates had proved to give some long-term security, and the competitiveness of windpower had increased.

Much of this newly installed windpower was, however, installed and financed by non-utility power producers. The large power companies started to develop (or buy) projects and got some large windpower plants on line from 2008 and later. In 2012, windpower was considered as mainstream technology in the power business. In March 2013, the branch organization *Svensk Vindenergi*, managed by the power companies in the DES, presented a plan for a transition to a 100 pecent renewable energy system in Sweden by 2030 (Svensk Vindenergi, 2013).

2.2.11 *Transition and ownership*

What is the relationship between the transition to a renewable power system and ownership of power plants?

The development has been described by many researchers referred to in this chapter as starting as a niche market and gradually, as technology is proven and becomes truly competitive, becoming accepted by the actors of the DES. The large power companies start to invest in and use windpower just like any other power plant. In most cases, however, this takes a very long time.

To implement this transition, it has been necessary to make it possible for other actors, such as farmers, non-utility companies, non-governmental organizations, and private consumers, to invest in and connect power plants to the power grid. This was made possible by political decisions.

Laws to make this possible were introduced in Denmark, Germany, and some other countries, and development started. Without this new legislation, no transition would have been possible. Without these early investors, the technical development that has made windpower competitive in the market would not have taken place.

The possibility of ownership by actors outside of the DES has been an important precondition for transition in many countries. This was the case already before the market was deregulated, so the deregulation was not a necessary condition for a transition to start. The feed-in laws, which gave the right to independent power producers to be connected to the grid, made this possible.

What will happen when windpower has become a mainstream technology, as has happened during the last few years? Will the actors from the DES take over the ownership of windpower plants and become a dominant player, or will these new independent owner categories continue to play a significant role?

The development of windpower on a global as well as a national Swedish level is described in Chapter 4, Windpower Development and Ownership. Before that, the electric power system, how it works from a technical as well as an economical point of view, will be described in Chapter 3, Windpower in the Electric Power System.

Notes

1 Load factor is a measure of how much of the capacity of a power plant is used. A high load factor (more full-load hours) makes a power plant (and the investment) more efficient.

2 The tragedy of the commons means that overexploitation of natural resources that are managed as commons is unavoidable.

3 These organizations have regularly published reports and articles in the press criticizing windpower, IVA claims in a report 2009 that *Half as Much Windpower is Enough* (Klefbom, 2009) and KVA professors warn against "windpower optimism" (Karlberg, 2009).

4 A mutually supportive interaction between "technology development" and "market development" leading to a progressive "industrial development" of both the quantity and quality of RES technologies (Lafferty & Ruud, 2008).
5 In Spain, development slowed down in 2012 owing to the severe financial crisis.
6 This is an updated version of a section with the same heading in the book *Developing Wind Power Projects* (Wizelius, 2007).
7 Full-load hours means the production of a power plant operated at full capacity. How long this takes depends on the capacity factor of a specific power plant, which depends on the wind resources at the site.
8 Banking: power generated a by private company at one time of the year can be used at another time (i.e., the utility grid can be used for "storage"). Wheeling is the right to use a utility grid to transmit power from a windpower plant to a factory elsewhere.

3 Windpower in the electric power system

Wind turbines are components of the electric power system, but their properties differ on essential points from other power plants. These properties, such as intermittency, are creating problems, or rather challenges, to electrical engineers responsible for the proper operation of electric power systems.

These properties and how windpower can be integrated in the electrical power system are described in the first part of this chapter. A more detailed background on the electric power system, with its technical, legal/political and economic subsystems, is presented in Appendix B, The Electric Power System.

After that comes a section about the electric power market (i.e., the economic part of the system). A more detailed description of how the Nord Pool market works is presented in Appendix C, The Nord Pool market. Finally there is a section on how this market is evaluated by different actors.

3.1 Properties of windpower plants

The electric power system is, from a technical point of view, complicated and complex. The development of this system has contributed immensely to the development of the modern industrial society and to the social well-being of humankind. The impacts on social development are immediate and important: a single electric light bulb in a rural village in a developing country will make it possible for children to do their homework and to become educated.

In the developed world, most people take electricity for granted and do not realize its role until there is a blackout. Those who operate the electric power system have a large responsibility, and it is understandable that electrical engineers are careful and rather conservative when it comes to the introduction of new artifacts, such as wind turbines, with some unknown or unusual properties, to this system.

Wind turbines are rather small, with an average nominal power of around 2 MW in 2010, compared to conventional power plants with nominal power from around 10 MW to more than 1,000 MW.

Windpower is a modular technology: windpower plants can consist of a single turbine, small groups of turbines, or large wind farms with hundreds of turbines. Single and smaller groups of wind turbines are connected to the distribution grid to provide local production for local consumption: *distributed generation.*

Availability differs as well. Power from wind turbines is available only when there is wind and varies very much with the wind speed, since the energy in the wind is proportional to the cube of the wind speed (Manwell, McGowan, & Rogers, 2002). This means that windpower can be used neither as base power nor regulating power nor peak power in the electric power system. Instead, the other components (power plants) have to adapt their production to the actual power production from wind turbines.

Because wind turbines convert energy in the wind to electric power, they do not need any fuel, thus no transports of fuel, and the costs are not dependent on the availability and price of raw materials on the world market. The wind is a local resource for local use (although it may be transmitted far in a power distribution system).

Among the renewables, windpower and solar energy have this property, while renewable fuels from biomass and the like are traded on a market and transported to power plants.

Windpower has these specific properties:

- moderate size (and price);
- modularity;
- intermittency;
- distributed generation;
- local resource utilization;
- independence of raw materials market (and price);
- no fuel transport; and
- no emissions to environment.

These properties define the role of windpower in the electric power system and in the global energy- and ecosystems.

3.2 Integration of windpower

How do these properties and their impact fit into the electric power system?

3.2.1 Moderate size and price

The moderate size and price of wind turbines open up the opportunity for new actors to enter the power system (economics), given that the laws allow independent power producers to connect to the electric power grid (legal/political system), which in turn will add wind turbines to the power system (technical system).

3.2.2 Modularity

The modularity—which means that a windpower plant may consist of anything between 1 and 1,000 wind turbines—adds flexibility to the power system, with shorter lead times for development than nuclear power plants; for example, it is easier to adapt power production capacity to a changing demand. To fill this role, however, the legal/political system must enable a fast process for granting permissions to such projects.

3.2.3 Variability

Variability (i.e., that the power production varies with the wind) will put new demands on grid operators as well as wind turbine manufacturers to enable a high windpower penetration in the power system (technical system). With higher penetration, wind turbines have to be able to contribute to the stability of the power system and participate in power regulation: to control voltage stability and frequency. Wind turbine manufacturers have thus developed technology to accomplish such demands (technical system).

The value of wind-generated electric power can be increased with good wind prognoses that could make it possible to predict the power production during coming hours or days with high accuracy. Prognoses will, of course, not have any impact on the production of the turbines, but it will have a considerable economic value. If predictions and actual production coincide, the penalties for power producers for deviations from "ordered" power will decrease.

3.2.4 Distributed generation

In the early days of windpower, when the turbines were quite small (20–250 kW), they were intended to produce local power that could be used in the close vicinity, and even today many turbines actually function in that way. This is an obvious advantage for the power system, as it will reduce the losses in the grid (the longer distance power is transmitted, the larger are the losses). As long as the maximum power from the turbines is less than the minimum load in the local grid, all power from the wind turbines will be consumed in the local area.

A factory with a large and relatively constant power demand can use all power from a wind turbine for its own use. Some factories that operate day and night all year round have their own wind turbines and feed the power directly into the factory where all the power will be used.

3.2.5 Local resource utilization

Wind turbines utilize local wind resources. The advantage is that no raw materials have to be transported to the site during its lifetime. This saves a lot of costs for transport vehicles and increases energy security.

3.2.6 *Independence of raw materials market (and price)*

As no raw materials such as coal, oil, gas, or biomass have to be used and the wind is free, the costs of production are predictable. The capital costs and costs of operation and maintenance are known in advance and regulated in contracts. This property is also an advantage for energy security and the balance of trade.

3.2.7 *No fuel transport or emissions to environment*

Both these properties are advantages for the environment and reduce emission of greenhouse gases (GHG) and hazardous emissions. As reduction of GHG emissions is on the political agenda (legal/political system), these properties give politicians a good motive to introduce support to windpower (economic system), which will add more wind turbines to the power system (technical system).

3.3 Technical acceptance

Some five or six years into the twenty-first century, windpower seems to have been definitely accepted as a component in large electric power systems by most operators in the world. To get a high windpower penetration, there are still some demands that have to be met, and additional equipment in the wind turbines will be necessary. This kind of technology for large-scale integration in the power system has, however, already been developed and implemented, for example in Spain, and will sooner or later be used also in the rest of the world.

The last technical barriers to windpower have been almost removed, and for the future there are rather local conditions, such as wind resources and the power mix, that will define the amount of windpower that can be integrated into electric power systems in different countries.

3.4 The electric power market

In the early days of the electric power system, there was fierce competition between different power companies; each had its own power grids, voltage levels, and the like. As the system grew, politicians found it necessary to bring order into this chaos and regulated the market. Electric power was turned into a *public service*, and prices were negotiated between power companies and politicians.

At that time, the supply and distribution of power were strictly national matters regulated by the state and considered a matter of public service and national security (Hughes, 1993). Now in Sweden and many other countries, the electric power market has been deregulated, and market places for electric power, such as Nord Pool in Scandinavia, have been created.

3.4.1 *Market deregulation*

During the early 1990s, the deregulation of the electric power markets was set on the political agenda in the European Union, the United States, Sweden and elsewhere (Högselius & Kaijser, 2007). Deregulation advocates argued for a radical change of policy. The first step would be to separate power production, distribution, and trade. Power plants would compete and get an incentive to cut costs that would press down prices for power. All power producers would get free access to the grid. Finally, electric power should be a commodity that was bought from power producers and sold to customers by independent power trading companies. The final step would be to open the boundaries and interconnect the power grids in different countries and regions to create an open international market for electric power.

The basic principles of a deregulated market are:

- production, distribution, and sale are separated, and no company may have more than one of these roles;
- every power producer has the right to connect to the power grid, if it is technically possible; and
- every consumer may choose any company to buy power from.

Grid operation is treated as a natural monopoly, and for each region a specific grid operator has an exclusive right to operate the power grid. These concessions are granted by a public agency (the state). A public agency (the state) controls whether these rules are adhered to and that grid tariffs are reasonable.

On the electric power market, electric power is a *commodity*, with no direct relation to the actual physical product: electric power (the oscillating electrons in the power grid).

In some countries, such as Sweden, Norway, the United Kingdom, and some states in the United States such as California, the markets have been completely deregulated. Other countries have just started with the deregulation, and there are countries that resist this process.

Two important questions that have to be solved in the deregulation process are who will have the responsibility for the overall system balance and how this can be maintained. In a power system, supply and demand have to balance at all times, otherwise the system will collapse.

The power crisis in California after deregulation, where the power prices skyrocketed and many huge power companies, such as Pacific Gas and Electric, went bankrupt, proved that deregulation is a tricky business. To avoid such situations, someone has to have the overall responsibility and the legal power to order producers to supply the power needed to balance demand.

The deregulation process will most likely continue and will work when rules and regulations have been properly adjusted. Whether this will lead to lower power prices for end users is, however, not self-evident.

In a public service system, the actual cost of production will set the price; in a market system, it is supply and demand. When demand increases, prices will increase independent of the cost for producing the power. When demand decreases, prices will fall. However, power companies also have the option to lower supply to keep prices on a higher level. The questions of whether the electric power market actually works, whether there is a free competition, and whether prices are fair, have been the matter of much public debate in Sweden (Högselius & Kaijser, 2007).

3.4.2 Deregulation in Sweden

In Sweden, a new electric power law came into force on the first of January 1996. The power market was deregulated, or re-regulated, as some prefer to say, since there still is a legal framework that regulates power production and the market. With this new law, the monopolies for sale of power were abolished, and power trade was opened to competition. Every customer can now freely choose a power company from which to buy electric power.

Production, trade, and distribution of power are strictly separated. These tasks used to be managed by the same company—municipal utilities, state-owned Vattenfall, or private power companies—that had reserved rights to sell and distribute power within their regions of operation.

According to the new law, trade with power shall be handled by power trading companies and the distribution of power by grid operators. Many power companies and utilities have created separate affiliated companies to handle the power trade and grid operation. These are companies within mother-companies: the requirement is that the bookkeeping is kept separate, so that no money can be moved from grid operation to power trade and so on.

Grid operation is still a monopoly, as it is considered too expensive and irrational to install competing power grids. The energy market inspectorate issues licenses to grid operators. These get sole rights to distribute power in specific regions. Prices for this service have to be reasonable and are supervised by the energy market inspectorate. The price for power is, however, completely free; it is regulated only by supply and demand on the power market.

A customer of electric power always has to make two separate contracts, one with a power trading company for the purchase of power and another with the grid operator that distributes the power to the end user. The customer pays one fee to the grid operator for the distribution of the power and one to the power trading company for the electric power. On top of this comes an electric power tax and green certificate fee. Finally, 25 percent value added tax on the sum of all these different entries is added.

The grid fee constitutes about 20 percent, and the power price and tax around 40 percent each of the electric power bills. The only cost that can be changed by choosing another power trading company is the price for the power itself, which constitutes less than half of the power bill.

3.4.3 The Swedish certificate system

In Sweden, a system with green certificates (in Sweden called *elcertifikat*: electric certificates) was introduced in 2003 to promote development of renewable energy. Renewable energy sources, such as windpower, biomass, hydropower, and so on, will compete with one another on equal terms, and the price is set by supply and demand on the certificate market. The cost for the certificates will be paid by consumers.

Energy-intensive industries are excluded from the obligation to buy certificates; however, they can earn and sell certificates if they have some renewable power production, even if the industry uses this power within their own factories.

Power plants that produce renewable electric power have to be approved by the Energy Agency.

Power plants that use these energy sources can register to get certificates:

- all wind turbines (independent of size);
- solar power from PV-cells;
- power plants using biomass as fuel;
- small-scale hydro power plants (< 1,500 kW);
- new hydro power plants;
- retrofits of existing hydro power plants; and
- CHP plants using peat[1] as fuel.

When these producers of renewable electric power have been approved and registered, they will get one certificate for each MWh/year (1,000 kWh/year) from the state. Certificates will be issued for fifteen years. The Swedish National Grid (Svenska Kraftnät) checks the production of the power plants (Swedish Energy Agency, 2010).

All power trading companies are obliged by law to have a specific share, *quota*, of renewable energy in the power they are selling to customers. They can obtain this share by buying certificates from power producers. The cost for these certificates is then passed on to the end users. The sizes of the quotas are decided by parliament and, to increase demand for renewable energy, the quotas are raised from year to year. If political targets are not met, the quotas can be modified to increase demand and thus certificate prices.

There is a penalty price for power trading companies that do not fulfill their quota obligations. Certificates are issued annually but do not have to be sold any specific year.

Norway was integrated in the green certificate system from the beginning of 2012.

3.4.4 Nord Pool: The Nordic electric power market

The Nordic countries—Sweden, Denmark, Norway, and Finland—have a common electric power system and also a common marketplace: Nord Pool, which is managed and operated by Nasdaq OMX.

There are six different actors on the market (Pöyru, 2010):

- *TSOs*: Owners/operators of the HV transmission network, with overall responsibility for the electric power system;
- *producers*: owners/operators of power plants;
- *traders*: companies that buy power from producers and sell it to consumers;
- *brokers*: companies that negotiate business deals between producers and traders;
- *consumers*: end users of electric power; and
- *grid operators*: measure and report power consumption.

How they interact is illustrated in Figure 3.1

The price on the Nord Pool spot market is set by the marginal cost of the most expensive power that has to be used to satisfy demand. Power producers submit bids for how much power they can deliver and at what price twenty-four hours ahead to Nord Pool.

The lowest bids, usually from hydropower and windpower with low marginal costs, are taken first and then the more expensive bids, until the demand is covered. All producers get the same price, corresponding to the highest bid for power that will be used (see Figure 3.2).

The power price depends on several factors. On the supply side, the short-term climate has a great impact; for hydropower, it is the annual perspiration, which varies significantly different years. For windpower, it is the real-time wind that matters, and many efforts are made to develop good wind prediction models. Fuel prices and, in Europe also, emission permits have an impact on the power price from thermal power plants.

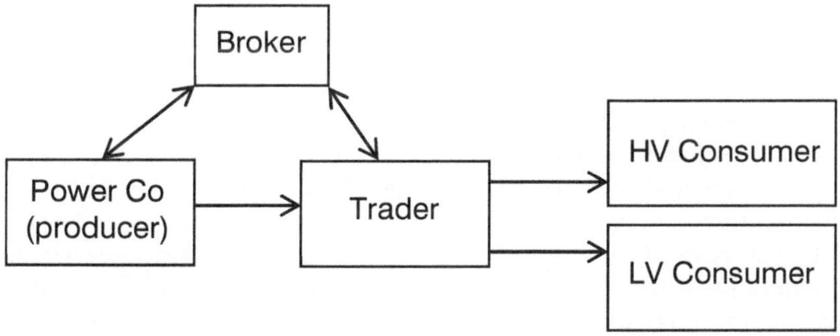

Figure 3.1 Model of the trading system. All power on the electric power market has to pass a trading company. The power producer sells the power to a trader, who sells it to the end user. Sometimes the prices are negotiated through a broker. The contracts can be based on the hourly spot price on Nord Pool, on monthly average price on Nord Pool, or on one- to three-year contracts often based on Nord Pool futures (expected price)

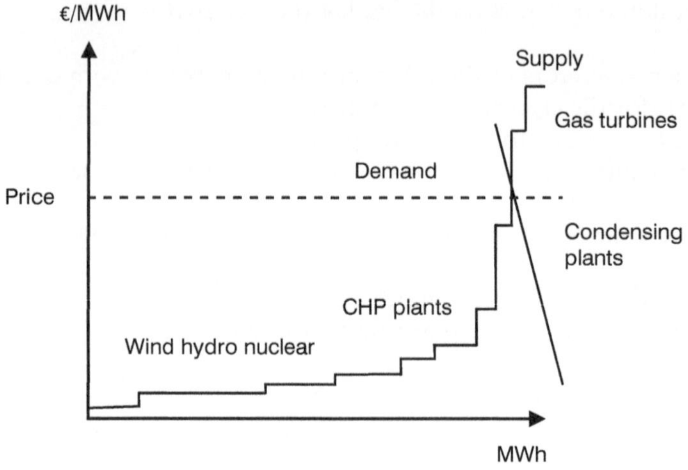

Figure 3.2 Supply and demand curve for electric power exchange. When demand increases, the power price will increase, too. This follows from the merit order system, where the power plants with lowest costs are used first (wind, hydro, and nuclear), and plants with higher costs (CHP plants, condensing plants, gas turbines) are added until supply equals demand. All producers will then get this price (dotted line) for the power produced and delivered (Pöyru, 2010)

On the demand side, it is also the weather, the temperature, and the winds that determine how big demand will be, especially when electric power is used for heating. The international economic situation and power demand from the industry play a big role as well. Power producers also have the option to sell the power on other markets if they have access to them. Power consumers do not have a similar option (see Figure 3.3).

3.4.5 Regulating power market

If the scheduled power production (supply) does not fit the actual power demand (predicted twenty-four hours in advance), the power production has to be regulated (increased or decreased). To handle this need there is a regulating power market.

Balance responsible players (BRPs; i.e., power companies or large consumers who have a balance agreement with a TSO) bid the regulating power to the regulating power market, and the TSO determines the use of reserves (and regulating power market) according to net imbalances of the (Nordic) system. After the operating hours, the net imbalances of all BRPs are calculated and charged for according to country-specific rules (Wangensteen, 2007).

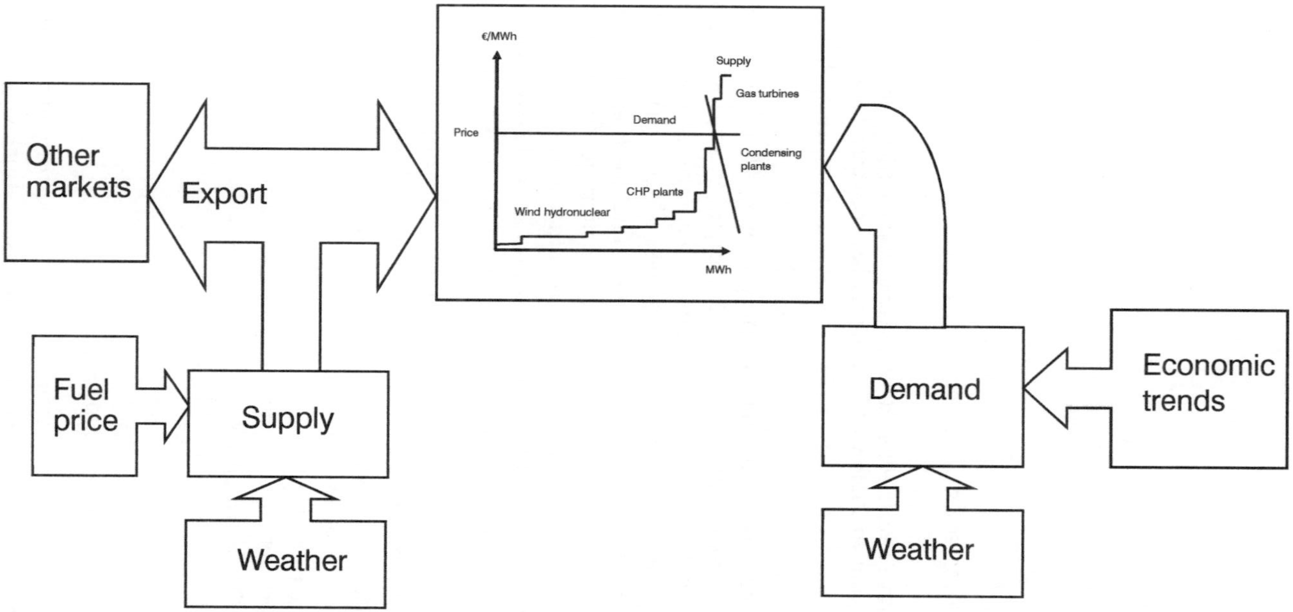

Figure 3.3 Factors with impact on Nord Pool electric power price. The demand for electric power varies according to regular diurnal, weekly, and seasonal patterns. On top of this, global economic trends and variations in weather conditions have impacts on demand. On the supply side, weather conditions have a direct impact on the production from wind turbines and solar cells and seasonal variations of hydro power. World market fuel prices as well as price levels on export markets for electric power from the Nordic countries have an impact on the supply to the Nord Pool market

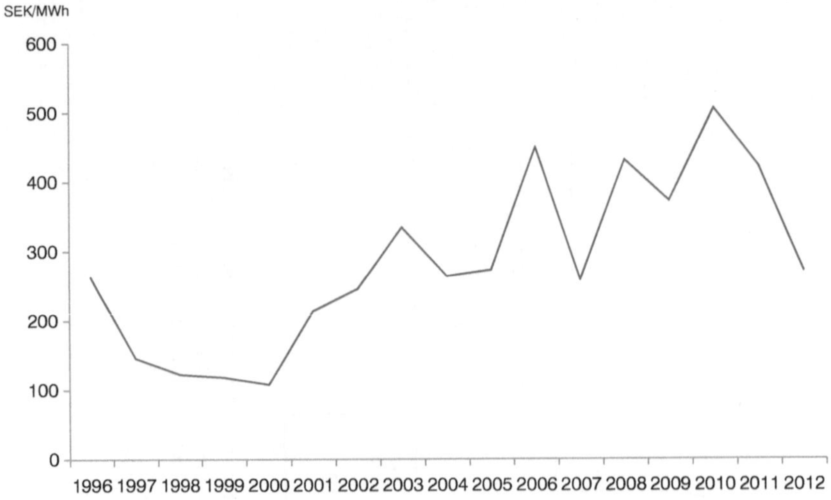

Figure 3.4 Nord Pool average prices, 1996–2012. Average annual prices have varied from 100–500 SEK/MWh, but volatility within each year has been even bigger

3.4.6 Market prices

Since deregulation, power prices on the electric power market have been rather unstable. The first years after deregulation, prices went down and stayed on a low level up to the year 2000. After that, prices have been increasing, but in 2012, there was a sharp decrease (see Figure 3.4).

The certificate prices have also been volatile. The justification for introducing the certificate system was to replace direct subsidies financed by the state budget (by taxes) with a market-based system that would be paid by the consumers.[2] However, the value of the certificates is set and governed by politicians, as the size of the quotas (demand) and maximum price (penalty) are decided by parliament.

The certificate system can, if it works, be used either to stimulate or to limit the development of renewable energy and windpower. The system has, since it was introduced, been revised several times, when the quotas for coming years have been changed to increase demand and prices. Still, prices have been very volatile (see Figure 3.5).

Windpower producers get their revenues from two markets with highly volatile prices. When the power prices and the certificate prices are low simultaneously, it is difficult to make ends meet. Investments in windpower have a high risk.

3.4.7 The impact of windpower on market prices

Although windpower needs some economic support to be competitive, the impact on power prices from windpower is the opposite; more windpower

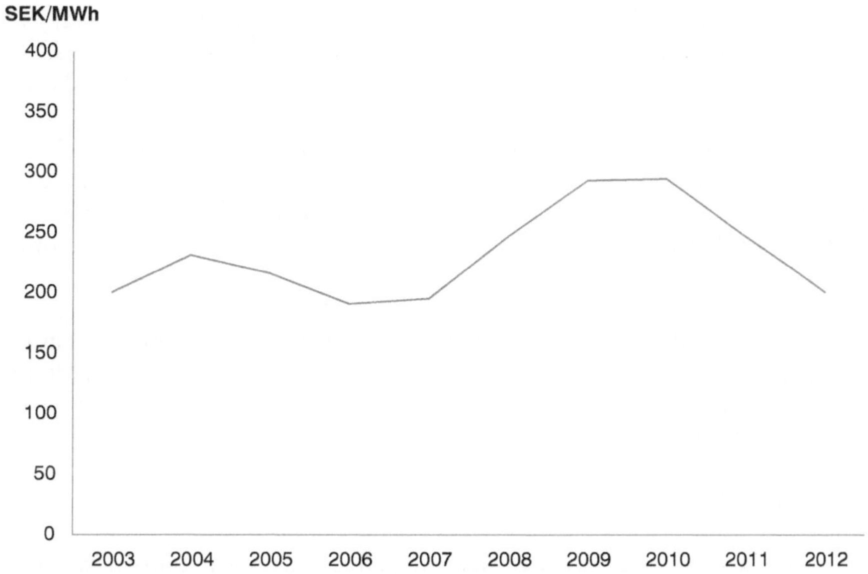

Figure 3.5 Prices for green certificates, 2003–2012, SEK/MWh. Average annual prices have varied from 200 to 300 SEK/MWh, but volatility within each year has been much bigger

in the power system will lower the spot prices on the power market. This is caused by the so-called merit-order-effect. This effect has been studied in many scientific papers that have been reviewed by the consultancy company Pöyru for the European Wind Energy Association (PÖYRU, 2010).

The conclusions are that an increased penetration of windpower will reduce wholesale spot prices and also that consumers will pay lower prices for electric power. According to another study, windpower also reduces intra-day variability of wholesale electricity prices on the Nord Pool spot market (Mauritzen, 2011).

3.4.8 Evaluations of the Nord Pool market

Ever since the electric power market was deregulated in 1996, the question whether this market really is a free market where competition between producers and traders will reduce power prices for consumers has been a topic for discussion in Sweden.

A government report published in 2002, *Competition on the Electric Power Market* (SOU 2002:7), starts out by defining the purpose of a competitive electric power market:

> The aim of the reform of the electric power market was to increase efficiency in production and sale by introducing freedom of choice for

> the consumers. Competition in the power trade should create conditions for an increased pressure on prices and costs in the power business.
>
> (Government Report, 2002, my translation)

Some concerns are raised; one is that the number of power producers has been reduced, another that power prices have tended to increase. The volatility of prices is mainly due to hydropower and the variations in perspiration between different years, according to the report. Three other factors are mentioned:

- *Energy balance*: The energy balance has declined, which means that the marginal cost for power from plants with fossils fuels in Denmark and Finland governs the price on the market.
- *Fuel prices*: The price of fossil fuels has increased.
- *Emission permits*: Costs for emission permits from these plants have been added.

Several actors have expressed suspicions that the power companies, which jointly own and operate the nuclear power plants and other large power plants as well, cooperate to reduce market competition. However, according to the report, there are few signs of price manipulations from power producers.

Another concern is that the number of power trading companies has been reduced. There are 130 power trade companies on the market, but many of these do not compete with one another, as they are partly or wholly owned by others or bound by contracts. The three largest power traders, Vattenfall, Fortum, and E.ON, control some 70 percent of the sale to customers.

The main conclusion is that the market, given the institutional framework that governs it, works fairly well. However, the reduction of actors on the market in production and trade increases the risk for market power.

The state agency Energy Markets inspectorate *report according to the EU directive for internal markets for electric power and natural gas* (Energimarknadsinspektionen, 2009) was published in 2009.

The report states that an increased volume of power is traded on Nord Pool; in 2008, it was 76 percent of the power traded in the Nordic countries. The number of power companies with more than 5 percent of the production in Sweden was reduced to three: Vattenfall, Fortum, and E.ON. Of the 124 power trading companies, only 20 were independent from these three mentioned power conglomerates. The conclusion is, however, that the market functions quite well.

In 2009, Petter Rönnborg published his PhD thesis, *Det där ordnar marknaden...* (in English: *The market takes care of that...*) (Rönnborg, 2009) at Gothenburg University. He made interviews with different actors on the electric power market, mainly power companies and municipal utilities, with a focus on how decisions about investments in new power plants were made.

The informants give a more controversial view on the benefits and the working of the deregulated power market. There is a big disappointment with the fact that power prices have increased, as the motivation for this reform was to get lower power prices by more competition. In fact, the power prices (including taxes, etc.) for industry have increased by 50 percent and for private consumers by 60 percent since 2002.

The increased prices have been described as a threat against the Swedish industry, and in the spring of 2006, representatives for the industry demanded a re-regulation of the market.

> "I said that we would regret a sale within ten years, but I was wrong, it took only two years," explains Christer Sjölin, former CEO of Stora Kraft AB, a power company that belonged to the paper pulp manufacturer Stora AB, when they sold their hydro power plants.
>
> (Rönnborg, 2009)

A CEO of a power company explains that "you don't have to cooperate on the market (with competitors), you know anyway how they will act."

A CEO from a municipal utility explains what happened with the deregulation: it was a change from producing power for a reasonable price for the citizens to optimize the profits of the company.

3.4.9 *From controlled monopoly to market oligopoly*

The electric power market can be described as an oligopoly market, with a few very large actors who control the market for their own benefit. Whether they actually manipulate prices is not an issue, as it is so simple and obvious how to keep the prices high by controlling production, export/import, and so on. In theory, it is a free market; in practice, it is not. The main reason for this is that the actors on the market are too few and are well informed on demand, and consumers have no option to respond to high prices as the power cannot be turned off.

The economy journalist Carolina Neurath from the main Swedish conservative newspaper *Svenska Dagbladet* characterizes the Nord Pool market as a "Donald Duck market":

> Imagine that the Stockholm stock exchange rushes 500 per cent in a few hours. And the responsible people calmly explain that everything is ok. This couldn't happen there. But on the Donald Duck market exchange Nord Pool, with wild trading and speculation on the electric power price, a commodity that has impacts for almost every citizen, this is allowed. But this market—which affects people who hardly know that this market exists and even less about how it works, seem not to need the same surveillance.
>
> (Neurath, 2010, my translation and summary)

Many letters to the editor have been criticizing power prices and the power market. One of many examples was published July 15, 2010, in the daily newspaper *Dagens Nyheter*:

> The high electric power prices during the first quarter of this year have had two reasons which partly are connected. First, power was produced by expensive fossil fuels, which decide the price for all electric power, also the cheap power from hydro power plants. As a result, Vattenfall made a profit of 7 billion (Swedish crowns) during the first quarter.
>
> This transfer of wealth from consumer to producer is caused by a system-error on the Nord Pool market, which many have pointed out. ....
>
> Second, the competition between the nuclear power plant owners Vattenfall, Fortum and E.ON has stopped after they have bought shares in each other's companies and their managing directors sit in all three companies' boards. They can cooperate and reduce power from the nuclear power plants to increase power prices on the Nord Pool market.
>
> This lack of competition is illegal, against EU-directives, and has to be stopped.
>
> (Setterbom, 2010, my translation and summary)

In February 2011, *Dagens Nyheter* had this headline in their economy section: *This is how the electric giants make their dream profits.* In the article, the journalist shows that it cost only an average twenty-four öre to produce 1 kWh of electric power, but if you want to buy the power, you may have to pay four times as much (Karlsson, 2011a).

This explains the billion-profits the power companies have been reporting during recent years and the shockingly high power bills consumers have had to pay, according to the reporter.

The Swedish Competition Authority has several times criticized the lack of competition on the electric power market.

> The producers have a position of strength and can easier control the market. They take no risk not to be able to sell what they produce. This is shown when they make their bid on the power exchange, according to Stig Arne Anker at the Competition Authority, who thinks that the power companies in this way keep prices high.
>
> (Quote from newspaper article above, my translation and summary)

Also Petter Rönnborg's PhD thesis got a lot of attention in the media. The economy section of *Dagens Nyheter* in March 2011 published an article with the headline *The system embeds for expensive power* (Karlsson , 2011a).

Rönnborg criticizes the politicians, power companies, and economists who, with such intensity, defend the Nord Pool power exchange, although it is obvious that the effect is giant profits for power producers and high prices for the consumers. As the market works today, there are no incentives for producers to compete by producing cheap power, he claims, and continues:

> To stubbornly defend a theoretical principle which may be nice and beautiful when written on paper is strange. You have to evaluate how the theory actually works in practice, see who wins and who will lose on this system.

Also the rising prices for grid operation have been scrutinized and criticized by the media. The increases have been very high where the grid is operated by the commercial power companies but significantly lower where the grid operator is a municipal utility (Villaägarna, 2004).

In a review of *When Public Power Became International*, (Högselius & Kaijser, 2007) in the culture section of *Dagens Nyheter* in July 2007 (Larsmo, 2007) the literature critic Ola Larsmo gives a good summary of the public view on the electric power system in Sweden at that time. Electrification has been the driver for the creation of modern society, he notes, and continues:

> In 2006 the storm Gudrun caused a severe blackout of the electric power system, which in some districts in Sweden lasted for weeks, and proved how vulnerable the modern society is for blackouts, and also focused on the lack of maintenance of the power grid in some parts of Sweden. These kinds of blackouts occurred again and again, in different parts of Sweden, for several consecutive years.
>
> Many outbursts of popular wrath have been caused by an ever more confusing, if not to say incomprehensible electric power market. It is not only anger over incompetence or strange business models. On a deeper level it is a worry that the modern society is at its end.
>
> (Larsmo, 2007, my translation)

Then a detailed account of what happened when the success story of Swedish power production was interrupted: Vattenfall was turned into a commercial company, and the politicians offloaded the responsibility for energy supply to the so-called market.

The retreat of politicians from public service, giving it over to commercial interests, is discussed by Colin Crouch in his book *Post-Democracy* (Crouch, 2004).

> Since the 1990s governments have increasingly renounced any distinctive role for public service, i.e., to provide citizens with equal services of a high standard, and require departments to act as firms, i.e., to provide services with a quality required to meet financial targets. Parts of public service is thus either privatized or contracted out to firms, or required to

> act as though they were firms. Government tries to divest itself from all direct responsibilities for the conduct of public services. And by doing so it relinquishes its claims to the special functions that can only be performed by public service.
>
> (Crouch, 2004)

This general development has in the electric power sector meant that by deregulation, electric power ceased to be a public service and was turned into a commercial commodity.

Vattenfall is an excellent example. This state-owned company did, after it was turned into a state-owned business company (where government appoints board members), act in direct opposition to the government's political aims and strategies.

According to the Swedish energy policy, no new nuclear power plants should be built. There is a goal to make a transition to a renewable energy system. To reduce emissions of carbon dioxide is a part of the climate policy. When the first Swedish nuclear reactors in Barsebäck were closed in 1999 and 2005, after a political decision, Vattenfall received financial compensation.[3] Vattenfall invested this money in nuclear power plants in Germany. Vattenfall also made some large investments in coal-fired power plants in northern Europe.

> A consequence of this has been a complete loss of confidence on the part of the public services. Nothing can be done without guidance from the business sector. Government is turned into an institutional idiot. From this follows the basic policy recommendation of contemporary economic orthodoxy: the state had best do nothing at all, beyond guaranteeing the freedom of the markets.
>
> (Crouch, 2004)

According to Crouch, the key concept to describe this development is *commercialization*, where the assumptions are that the quality and cost efficiency of public service will improve if existing practices are replaced by commercial practice.

> By the mid of the twentieth century people acquired right to public services by virtue of their status as citizens, not because they could buy it on the market. It was a mark of democracy. Provision of these services through a market would demean their citizen quality.
>
> (Crouch, 2004)

Further, according to Crouch:

> Post-industrial capitalism has therefore started to try to undo the deals made by its industrial predecessor and tear down the barriers to

> commercialization and commoditization imposed by mid-twentieth century concepts of citizenship.
>
> (Crouch, 2004)

In Sweden the large power companies have gradually changed character from politically governed "authorities" (providing electric power at low prices as a public service) to business companies whose sole ambition and aim is to optimize their profits.

3.4.10 Critical evaluations of the certificate system

In an article in the economy section of *Dagens Nyheter* (Kaintz Rognerud, 2010), the main morning daily paper in Sweden, the headline is *Failed effort—the government's system with certificates has not given desired effect*. The article then describes the situation on the certificate market, with sinking prices. The reason for this is that there is a surplus of certificates, as the energy-intensive industry, which is exempted from the certificate obligation, has exploited the system by converting existing power plants (often within their factories) from coal or oil to biomass. The quotas, which define demand, have been too low, so that supply is greater than demand, which leads to lower prices. This makes windpower and other renewables less profitable, which delays investments.

3.5 Summary and conclusions

To summarize, there is a lack of trust both from consumers, economists, and industry and from the power branch in the power market as well as the green certificate system. So far, this has received some comments from politicians, but no political parties have any ambition to change any of these systems.

From those who promote and develop renewable energy sources, such as the trade organizations Svensk Vindkraftförening (Swedish Windpower Association) and Svensk Vindenergi (Swedish Wind Energy), these systems are not criticized either, as high power prices will promote investments in new power plants. A change of system also creates insecurity, which hampers investments. However, in April 2012 the Swedish Windpower Association sent an open letter to the government, demanding an increase of the quotas, to raise the certificate prices, as many windpower owners could not make ends meet when power prices and certificate prices were at such low levels as they were that year (Lindahl, 2012).

Although no one, except maybe the large power companies, is fully satisfied with the electric power market, politicians from all parties as well as actors in the energy business seem to be prepared to accept the current rules of the game. One reason for this may be that the state-owned power company Vattenfall is one of the winners and delivers its large profits to the state.

Notes

1 Peat was added afterward (April 2004) owing to strong lobbying from peat producers.
2 Feed-in tariffs are also paid by the consumers.
3 Barsebäck was owned by E.ON, but as a part of the compensation, E.ON received a share of Vattenfall's nuclear power plant in Ringhals, and Vattenfall received financial compensation from the state.

4 Windpower development and ownership

Windpower has developed very rapidly during the last thirty years. In the late 1970s and early 1980s, wind turbines were small, manufactured in smithies and small workshops, and installed next to farms. Thirty years later, wind farms on land and offshore are parts of the dominating energy system and are manufactured by industrial giants in the electric power business such as General Electric in the United States, Siemens in Germany, and their counterparts in India and China. In 2012, around 3 percent of all electric power in the world was produced by wind turbines (WWEA, 2013), a share that will grow for many years ahead.

4.1 Wind turbines for grid connection

The power in the wind has been utilized historically first by sailing ships and for work by different kinds of windmills for thousands of years (Hills, 1996). Wind turbines for electric power production were developed in Denmark by Paul laCour, and the first commercial grid-connected wind turbine started operating in Askov, Denmark, in 1902 (Nissen, 2003). From the 1920s, battery-charging wind turbines were in common use in the countryside in the United States until the power grid was extended to these areas some decades later. These kinds of small wind turbines were quite common also in Europe (Hills, 1996).

In the 1950s a wind turbine that delivered power to the AC-grid was developed, the so-called Gedser-möllan, which gave valuable experience to Danish researchers and electrical engineers when it was in operation in the 1950s. The abundance of fossil fuels that flooded the market at that time made it difficult for windpower to be competitive.

The modern windpower industry was developed in Denmark starting in the late 1970s and has been growing ever since. This was the start of the Danish success story in the windpower business, which some forty years later had spread across the globe.

Figure 4.1 Typical wind turbine from early 1980s. A Vestas V15 with 35 kW nominal power and 18-m hub height installed at Lövsta agricultural school on Gotland. Photo: Tore Wizelius

4.1.1 Ever-growing wind turbines

Commercial wind turbines for grid connection, which are the focus of this study, were rather small when development started in the late 1970s and early 1980s. At that time, the typical wind turbines had hub heights of about twenty meters, and on these lattice towers, there was a nacelle with a 20 to 30 kW generator, a rotor with a diameter of around 15 meters, and a swept area of 175 m^2 (see Figure 4.1).

In 2010, the largest wind turbines had hub heights of up to 160 meters, nominal power up to 7.5 MW, and rotor diameters close to 130 meters (see Figure 4.2) (Earnest & Wizelius, 2011). The industry was then developing wind turbines with a nominal power of 10 to 15 MW (Quilter, 2011).

From the start in the late 1970s, wind turbine size has doubled on average each three to four years, if size is measured as the most common nominal power of the turbines during a specific year. When nominal power is increased, the hub heights and the rotor sizes increase as well (see Figure 4.3).

4.1.2 Early starters and late bloomers

Denmark, this rather small country in northern Europe, has been at the forefront of the technical development of wind turbines as well as the actual installation of windpower connected to the power grid. A historical overview of the amount of installed capacity of windpower in the electric power system shows that Denmark was in the lead from the 1970s and up to the early 1990s, except during the short windpower boom in California in the 1980s.

In 1994, Germany passed Denmark and kept the number one position for a decade, with Spain as a runner up in the early years of the twenty-first

Figure 4.2 4.2 Modern large wind turbines. Gamesa G128 with a nominal power of 4.5 MW, 120-meter hub height and 128-meter rotor diameter, is one of the new very large wind turbines on the market

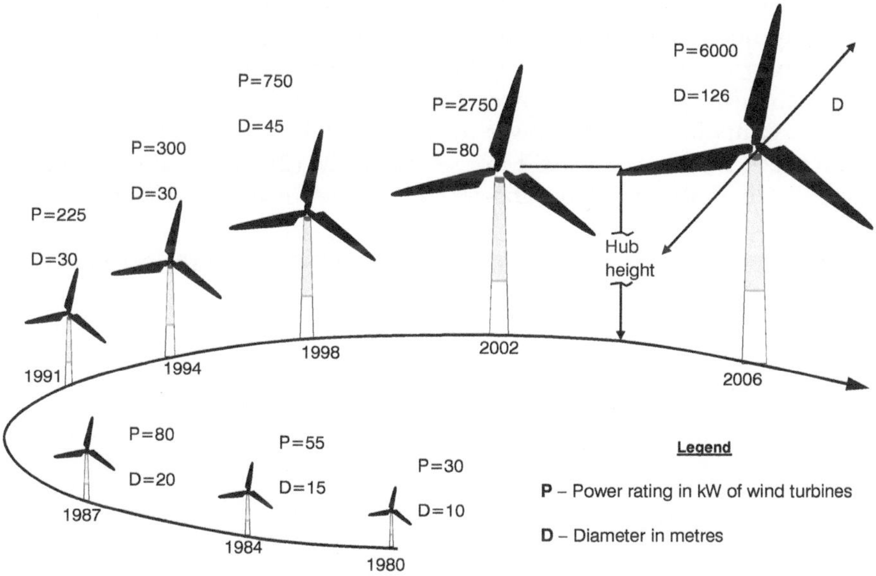

Figure 4.3 Development of wind turbine size. The size of wind turbines has increased rapidly, with a doubling every three to four years (Earnest & Wizelius, 2011)

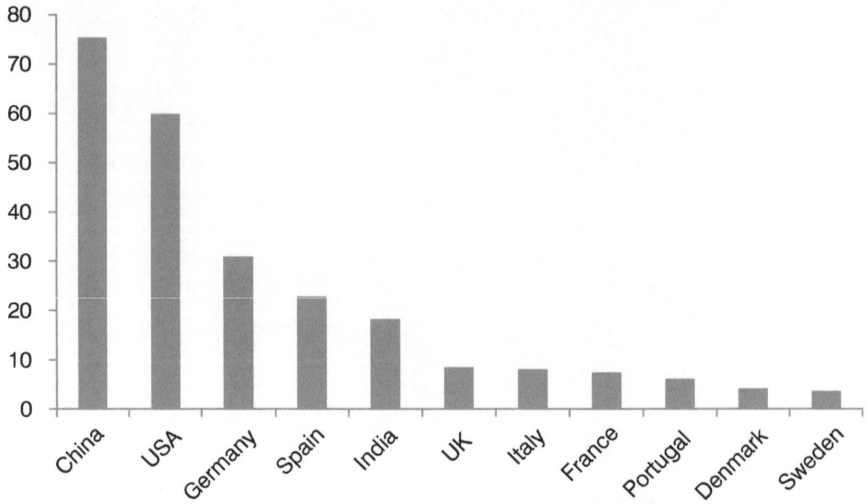

Figure 4.4 Top twelve countries by installed capacity in 2012, GW (WWEA, 2013)

century, the United States taking the lead in 2007, and China, with an annual growth rate of more than 100 percent a year for several consecutive years, taking over the lead position in 2010 (see Figure 4.4).

It may not be fair to compare a small country such as Denmark with giant countries such as India, China, and the United States. A better measure for comparison is the windpower *penetration*, which means the share of electricity supplied by windpower to the national electric power system. In this respect, Denmark was still in the lead in 2012. China and the United States, with most installed capacity, are below Sweden on this list (see Table 4.1).

Other measures are windpower per capita and windpower per land area in different countries (see Figures 4.5 and 4.6).

Table 4.1 Windpower penetration 2012, different countries

Country	*Penetration, %*
Denmark	27
Portugal	17
Spain	16
Ireland	13
Germany	11
Sweden	4.4

Source: (EWEA, 2013) (Energimyndigheten, 2013B)

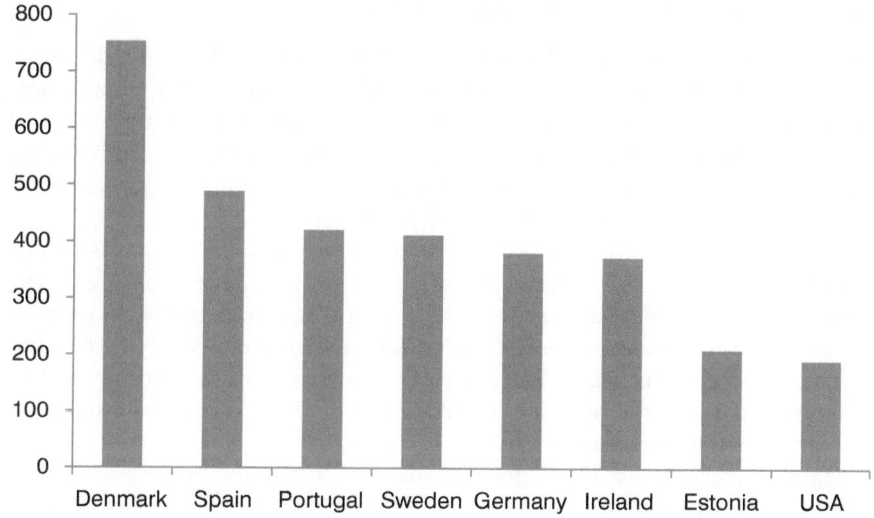

Figure 4.5 Windpower capacity per capita, 2012 (W/cap). In relation to the population, Denmark is still far ahead of other countries. With this measure, Sweden is in position number four (WWEA, 2013)

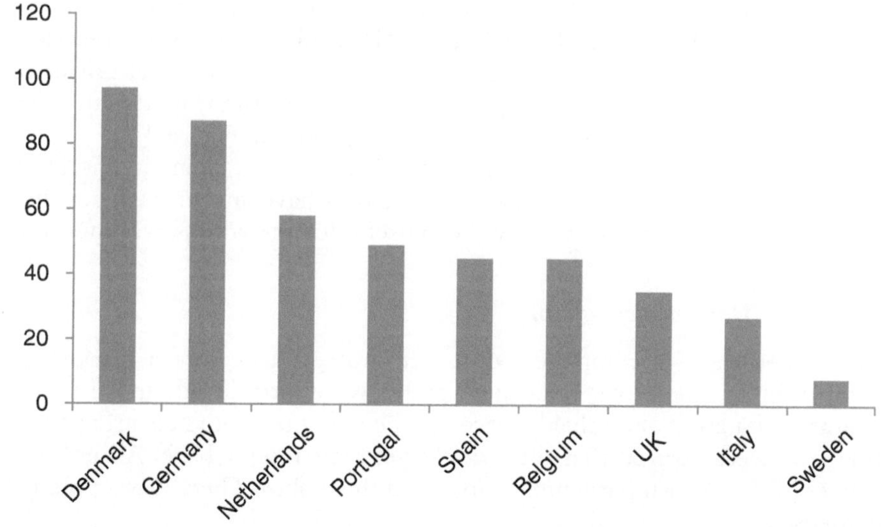

Figure 4.6 Windpower capacity per land area, 2012 (kW/km^2). In relation to land area, Denmark is also in the lead, but Germany is close behind. With this measure, Sweden is far behind. This shows that there is plenty of space for more windpower development in most countries of the world (WWEA, 2013)

4.1.3 Development of the windpower industry

The development of the windpower industry started when the oil crises during the 1970s created panic in the industrialized world, which was enhanced in the 1980s by the nuclear reactor accidents in Harrisburg and Chernobyl. Politicians as well as the public were looking for new energy sources. The option was to develop local renewable energy sources: solar energy, biomass fuels, hydropower, and windpower.

In the 1990s, the focus was moved to the ever-increasing emissions of carbon dioxide and other greenhouse gases that threatened to change the global climate. The burning of fossil fuels for power production was the main source of these emissions; energy policy and environment policy turned out to have a close connection. Since the UN Environment conference in Rio de Janeiro in 1992, the necessity to change to renewable energy sources has been given an ever-stronger weight.

Many governments then decided to invest money in research and development and entrusted the task to develop windpower to large power companies and the aircraft industry, as they should have the competence and the resources to succeed.

Sweden, Germany, the United Kingdom, the Netherlands, the United States, and many other countries started national windpower research programs. The aim was to develop large wind turbines with a nominal power of several MW, which was considered necessary for windpower to make a significant contribution to the power production on a national level.

During the 1980s, several very large R&D wind turbines were erected in different countries. In Sweden, large prototypes with a nominal power of 2 to 3 MW, Näsudden I and Maglarp, were built. In the United States and Germany, wind turbines of this size were built, too (see Figure 4.7).

These behemoths, as the author Paul Gipe named them in his book *Wind Energy Comes of Age* (Gipe, 1995), didn't have any success, but they provided some experience that has been useful for the windpower industry.

4.1.4 The Danish success story

The Danes chose a completely different strategy: to create a market for wind turbines. Generous investment grants were introduced, and the state guaranteed a good and reliable price for the wind-generated electric power. Investors were assured that they would get their money back. All political parties in the Danish parliament supported this policy. There was a political consensus.

In Denmark, a completely different kind of company started to manufacture wind turbines: small and medium-size enterprises, workshops, and smithies that produced plows and other farming equipment. They manufactured small wind turbines, 20 to 100 kW for the farmers' market (see Figure 4.1).

Figure 4.7 The end for the prototype Näsudden in Sweden. In June 2008, the Swedish 3-MW prototype at Näsudden was felled to the ground to make space for a modern wind turbine. An industrial heritage site was blown up. In a hundred years, this could have been a monument the effort to avoid global climate change. Photo: Ivo Palu

Farmers want machines that are reliable and robust, stand all kinds of weather conditions and, if they break, should be possible to repair in the farm workshop. The Danish companies manufactured wind turbines according to these principles. They did not opt for high-tech but for simplicity and durability. The first models were 15 to 20 meters high and with a nominal power of 20 to 30 kW, which every year became a little bigger. These companies had a large and faithful group of customers, a solid market; wind turbines became just another machine among their products.

Many farmers bought their own wind turbines, even though it was a new technology with considerable financial risk. Members of the public, worried about the environmental crisis, formed windpower cooperatives.

The windpower research center Risö tested and certified the wind turbines. Manufacturers did get feedback from the researchers, who had many good ideas about how the wind turbines could be further developed. Simultaneously, the number of wind turbines on line grew, and researchers and manufacturers got valuable feedback from the practical operation of hundreds of wind turbines.

In the early 1980s, the wind turbines had a nominal power of about 20 kW; by the year 2000, the size was approaching 2,000 kW. In twenty years, the size of the wind turbines increased by a factor of 100! The strategy to create a market for commercial windpower by investment grants and other subsidies in the 1980s turned out to be a much more successful strategy than large investments in high-tech research and development projects.

4.1.5 Germany and Spain catch up with the Danes

Germany introduced a 100 MW program for windpower development that soon was extended to a 250 MW program. The Danes got a new fast-growing export market. However, a large share of the support went to German manufacturers that started to compete with the Danes. In Germany and Denmark, economic support for windpower was seen not only as a part of the energy and environment policy but as part of the industrial policy; the support will make the manufacturing companies grow, so that new jobs are created and economic growth will increase.

From the year 2000, Spain has had a remarkable growth in the windpower sector, and the Spanish power companies play an active role in this development. Also in Spain, the political will to develop windpower on a large scale is not only a part of energy policy but of industrial and regional policy as well. Many old shipyards in the northwestern Atlantic coast have been taken over by windpower companies, and a lot of new jobs have been created in this once-depressed region of Spain. A domestic wind turbine-manufacturing industry was developed as well.

4.1.6 The Asian pioneer India and the boom in China

India has played a very prominent role as a pioneer for windpower in Asia. In the 1990s, several European manufacturers established their own factories or joint ventures, and domestic companies also began to manufacture wind turbines. Today the most successful Indian company, Suzlon, competes with European manufacturers on the global market (Earnest & Wizelius, 2011).

In China, a similar development started a decade later. China was a late starter in the windpower business, compared with Europe and its Asian neighbor India. In 1995, only 38 MW were installed, in a few small demonstration projects and with modest annual development in the range of 50 to 100 MW up to 2003. In 2004, 197 MW were installed, and the windpower capacity in 2004/05 added up to just 764 MW.

In the following years, a tremendous growth started, with growth rates in the range of 100 percent for several consecutive years: 503 MW (2004), 1,337 MW (2005), 3,304 (2006), 6,246 (2007), 12,209 (2008), and 25,810 (2009), adding up to a total of 75,564 MW in 2012–2013, moving China up to number one in the world. This rapid growth seems set to continue for many more years. Windpower has become a success story for China.

This rapid market growth had so far mainly been concentrated in just a few countries: Denmark, Germany, Spain, the United States, India, and China. In 2007–2008, the development took off in countries such as Canada, Australia, Brazil, France, and Portugal. Today, windpower is utilized—on a small or large scale—on all continents and in 100 countries and regions in the world (WWEA, 2013). Countries where development so far has hardly taken off will have a rapid market growth in coming years.

Commercial wind turbines are now bigger than the prototype behemoths that were built with huge R&D funds in the 1980s. There is, however, one important difference: the commercial wind turbines of today are competitive and reliable; they work. And for each increase in size, the cost per kWh produced has decreased.

During the same period, the projects have also grown larger. From installations with single wind turbines on farms in the 1980s and groups of three to ten wind turbines and even larger wind farms in the 1990s, the focus in Europe from 2000 and onward has been shifted to the development of large offshore wind farms, with big power companies as investors. Today, windpower is a major industrial branch in many countries and has become big business.

4.1.7 Supply and demand

From 2005 to 2008, demand for wind turbines was much larger than the supply. Developers who ordered wind turbines for their projects had to wait for years to get them delivered. It was also hard to place orders for single wind turbines—big orders have priority. And with a seller's market, prices for wind turbines increased. All leading manufacturers as well as subcontractors that manufacture components were expanding their factories as fast as they could.

After a period with mergers, when a few large and well-consolidated manufacturers seemed to get control of the market, and giant companies from the electric power business such as General Electric and Siemens started to manufacture wind turbines, this rapidly increasing demand created an unexpected but welcome opportunity for new manufacturers to establish themselves. In the last few years, many new companies have appeared, such as Bard in Germany, Leitwind in Italy, Goldwind and Sinovel in China, to name just a few.

In China, where windpower started to grow very rapidly in 2005, there are now around twenty new manufacturers of commercial wind turbines. New manufacturers and factories have also been established in Argentina, Canada, Korea, and many other countries. These new companies have not started by inventing the wheel once more. The know-how has been achieved through licenses from engineering companies or from joint ventures.

From 2011, there has been a better supply-and-demand balance. The global financial crisis in 2008 led to the cancellation of some very large projects, mainly in the United States, but the market has recovered. Some manufacturers in Asia are also starting to market their wind turbines, at very competitive prices, on export markets. Wind turbines are getting larger and more cost-efficient, and the competiveness of windpower is increasing.

4.1.8 Offshore windpower

Denmark was the pioneer also of offshore windpower. This rather small country had made a plan for windpower development in the early 1980s, and, according to this plan, offshore installations should have been built in the early 1990s.The first offshore wind farm was installed at Vindeby, and a few years later a second one at Tunö knob. Based on these experiences, a plan was made to install several very large offshore wind farms so that Denmark could get 50 percent of its electric power from the wind by 2030. The reason for this was that available land for wind farms on shore was limited and that the wind resources at sea were far better than on land. The cost to install windpower offshore is 40 percent to 50 percent more expensive, but the increase of the production will be on the same level.

During the last years, several offshore wind farms have been built: Horns Rev with eighty wind turbines of 2 MW each, and Nysted that is of the same size, and both have been expanded with a second phase of development. Several offshore installations have also been built in Sweden, the Netherlands, the United Kingdom, and Germany. In 2012–2013, Great Britain had by far most offshore windpower plants on line—2,950 MW—which was more than half of all offshore windpower (5,410 MW) at that time (GWEC, 2013). Many countries in Europe have started a large-scale utilization of windpower offshore, with installations that can produce as much power as a nuclear power plant. The most recent 5- to 6-MW wind turbines have been developed mainly for the offshore market.

4.1.9 Re-powering

Although windpower is a new branch, many of the wind turbines installed in the 1980s and early 1990s are getting old and need to be replaced. This process is called *re-powering*. By replacing groups of small wind turbines at good sites by fewer and larger ones, production can increase significantly. On the Swedish island of Gotland, six 3-MW turbines have replaced seventeen old wind turbines (150- and 500-kW wind turbines). This reduced the number of turbines by two-thirds, but these six new ones produce four times more energy. Similar re-powering schemes have been conducted also on the island Fehmarn in Germany and several other sites and will increase in coming years.

4.1.10 Phases of windpower development

When a new technology such as windpower is developed, it will pass through different phases. From the start in the late 1970s to 2012, the wind turbines have grown from small back-yard turbines for farms, sold on a local niche-market, to giant power plants with advanced power electronic equipment for grid integration manufactured in large factories on all continents and sold on the world market.

During this development, their role in the socio-technical system has changed as well. The ownership structure will change, and new ownership models will be developed. These structures and phases will also differ in different countries, owing to differences in the national characteristics of the electric power socio-technical systems.

Karl Sperling divides the development of windpower in pioneering Denmark into two phases (Sperling, 2007). The first phase starts around 1975, after the oil price shocks, and ends around 1996, when windpower had developed into a reliable niche technology. The goal during this phase was to develop windpower into a proven technology for a national wind turbine industry. It was characterized by incremental innovations, where bottom-up, trial-and-error activities of actors played a major role (Kamp, 2004).

This development was based on interaction and knowledge transfer between turbine manufacturers, turbine owners, and researchers, which enabled a participatory rather than competitive development of windpower technology (Kamp, 2004). On the local level, farmers became "owners-users." They established windpower businesses and organized meetings where knowledge was shared (Garud & Karnoe, 2003).

When the size of wind turbines and the investment costs had increased, local cooperatives were established. Most turbines installed between 1984 and 1994 were owned by cooperatives (Dansk Vindmölleförening, 2002). In 1996, windpower penetration in Denmark had reached 3.5 percent, and 120,000 individuals were part-owners of 4,000 wind turbines (Hvelpelund, 2006).

In 1996, the second phase of windpower development started in Denmark, characterized by large-scale integration of windpower in the power system. After that, windpower has become a major energy source. According to Sperling, it has, however, also reached a scale where political and economic resistance increased. Windpower now starts to become a business risk for conventional power companies, and in certain weather conditions may generate a significant electric power surplus that has to be sold at cheap prices at the Nord Pool market (Hvelpelund, 2006).

During the period 1996–2003, windpower capacity in Denmark increased very rapidly, from 600 to 3,116 MW (Energistyrelsen, 2011). From 2004, windpower installations in Denmark have stagnated, owing to local opposition and difficulties in finding new inland sites as well as reduced financial incentives. In 2010, the total installed capacity was 3,800 MW (Energistyrelsen, 2011). Most of the added capacity came from several large offshore wind farms owned by power companies. This has also led to a change in the ownership structure, from cooperative to corporate ownership.

Added to the two phases described by Karl Sperling (his report was published in 2007), the development from 2004 could be described as a third stage, with large offshore wind farm development, and re-powering schemes on land.

4.1.11 Market acceptance

A few years into the twenty-first century, windpower was no longer viewed as an alternative energy source doomed to a role as a marginal power producer by the large power companies in the dominant energy system (DES). It had become mainstream technology for the power industry. Some of the large multinational electric power equipment manufacturers such as General Electric and Siemens, which were established and played a leading role when the electric power system was created and expanded during the last century, started to manufacture and sell wind turbines. Large power companies from the DES, such as Vattenfall, E.ON, Iberdrola, and Enel, also started to invest in windpower plants on shore as well as offshore.

Some other new players entered the market as well: large international independent power producers (IPPs) such as International Power (UK), Invenergy (U.S.), and Acciona Energia (Spain). This group is defined as companies that have no grid operation and only invest in, build, and operate independent power plants (May, 2008). The consultancy company EER—Emerging Energy Research—confirms this development on its Website:

> Wind energy's emergence as a mainstream generation technology option has produced a dynamic global competitive environment with nearly 40 GW in annual capacity additions. Players across the value chain seek to scale their order books and project portfolios riding on wind's explosive 25% annual growth to capture a piece of this $40 billion market.
>
> (Emerging Energy Research, 2011)

The entry of these new players, power companies, and multinational IPPs and the ever-larger wind turbines pose new challenges to the power system and for acceptance of this technology. Will local ownership of windpower plants survive during this third phase of development?

Increased capacity means increasing investments by public and private investors. Larger wind turbines also make the necessary investments higher (but at the same time more cost-efficient). Most offshore windpower projects are also more expensive than sites on land, and are given more favorable economic conditions and, in the end, higher power prices for customers. The question is whether these investments are too large for some actors and will change the ownership structure.

During the first phase of windpower development, local ownership by cooperatives and limited companies played an important role for the motivation of local communities. The ever-larger turbines will likely increase local impacts and make it more difficult to get local acceptance. Local involvement and local benefits could be necessary also for windpower development in the future.

According to Szarka (Szarka, 2006), the technological state of the art of wind turbines requires large-scale investments that no longer can be realized through small-scale financing and ownership. This pessimistic view has so far proved to be wrong; there are small-scale financing options also for very large projects. Project developers have been very creative in finding new ownership, investment, and financing models. The flip-model in the United

States (Bolinger, 2011B) is one excellent example, and some examples from Sweden will be given in coming chapters.

Another very interesting development is the law on purchase rights in Denmark introduced on January 1, 2009, which guarantees people living close to new windpower plants the option to buy 20 percent of the wind farm at cost price. Such an offer has first to be made to people living within 4.5 km from the windpower plant, and if these do not fill up the 20 percent ownership, the additional shares are offered to citizens of the municipality where the wind farm is installed (Danish Government, 2008).

4.2 Ownership

The electric power system has developed from small isolated island-systems to a very large system where countries and even continents are interconnected. This system has to be coordinated at all times, supervised and controlled, so it has been considered to be a natural monopoly. Electric power has also been considered a national interest for national security and economic development. In the beginning, it was developed by private entrepreneurs and companies. When the power systems grew to large interconnected systems on a national level, it was in most countries also natural that it should be publicly owned. However, this applied mainly to the national transmission grids.

In many countries, there has been a mix of private and public ownership of power plants and local distribution grids. Until the 1990s, these systems with mixed ownership were strictly regulated by the state. At that time, a trend started toward deregulation and creation of power markets to set prices. This also opened the system for new entrants: independent power producers. Renewable energy technologies, mainly windpower plants, were modular and could be financed by new groups of actors, such as farmers and cooperative associations. Here is a short overview of ownership models in different times and countries.

4.2.1 Ownership and politics

In the books written by techno-historians, Hughes (Hughes, 1993) and Högselius/Kaijser (Högselius & Kaijser, 2007), which are more descriptive than theoretical, ownership of the companies involved in the development and operation of the electric power system are described in great detail. Hughes comparative study also discusses and analyzes the relevance of ownership, differences between municipal utilities and private power companies, as well as cooperation and competition between public and private sectors.

In *Networks of Power* (Hughes, 1993), Hughes gives a detailed description of the evolution of the electric power system in three major cities: Berlin, Chicago, and London. In this context, he also describes and discusses ownership models and the interaction of technology, economics, and politics during the period to 1930, when the electric power systems were fully developed in these large cities.

In Berlin, which was called *Elektropolis* due to the fast and efficient electrificationand the development of the large manufacturing companies Siemens and AEG, electrification (from 1880 to 1920) was characterized by an efficient coordination of technology and politics. The city government regulated the private utility, Berliner Electricitäts-Werke (BEW), that provided light and power to the city in a strict but also constructive way. To develop an electric power system is a very capital-intensive venture, and the private industry, BEW, AEG, Siemens, had very good contacts with financiers and banks.

The city, on the other hand, exercised a strong regulatory authority, and the contracts negotiated between the city and the utility (which got exclusive rights to distribute electric power) gave the city a share in the revenues of the utility; directed the services to satisfy the demands of households, electric transportation systems, and the growing industry; and had differentiated rates for these groups of consumers. There was also an option for the city to buy the utility, at specific economic terms, which the city finally did in 1915, when it turned into a public utility. At that time, all the staff, from management to workers, stayed with the company (Hughes, 1993).

In Chicago, the electric power system also developed and grew very quickly, but there the technology was the dominating factor. The electric power system developed relatively free from conservative constraints. Chicago was not known for the honesty of its government.

The strong entrepreneur and system builder Samuel Insull managed the Chicago Edison company, which grew to completely dominate the electric power system in Chicago and later large parts of the United States. Insull lobbied and finally also managed to move the management and regulation of the electric power system to the federal level, which was necessary to operate the system in a cost-efficient way. The utilities in Chicago were all private enterprises.

In London, the electric power system developed much more slowly; compared to Berlin and Chicago, it was very backward. While Chicago and Berlin had centralized power systems with large power plants, London had sixty-five local utilities, forty-nine different types of distribution systems, ten different frequencies, and thirty-two different voltage levels. It was not very efficient and was a nightmare for suppliers of electrical equipment.

The price of electric power was also very high, compared to Berlin and Chicago. London was at that time the largest city in the world with some 7.25 million inhabitants (1913) and had an extremely complicated administrative structure, reinforced by a strong respect for tradition. This affected the growth of the electrical system there, according to Hughes.

The administrative county of London was made up of the city plus twenty-eight other municipal boroughs, which all had their own systems. Local politicians also wanted to protect public investments in gas lamps for street lighting, and so on. It was very complicated and time-consuming to create new laws, and this could not keep pace with technical development. Electric power companies had difficulties getting approvals from authorities. The local authorities guarded their privileges in the face of would-be power suppliers. A centralized unified supply system was not developed until the start of the First World War. Some utilities were private, others publicly owned (Hughes, 1993).

To the electrical engineers and managers, a progressive urban utility was a mass-producing monopoly that supplied a universal system with a diverse load over a wide area at low rates and gave a fair return due to efficient production and maximum utilization of capital at reasonable interest rates.

When comparing these three cities, Hughes concludes that the backwardness of London was due to lack of large industrial enterprises and electrical equipment manufacturers. There were both private and local-authority-owned utilities in London, and their performance was quite similar, when measured on costs and return on investment. Ownership did not explain the backwardness; it was rather politics, and the laws.

In Chicago, there was a political go-ahead. In Berlin, there were some conflicts between technology and politics, but these conflicts were negotiated and did not create a stalemate. In London, they were not; there was confrontation between local government and utilities and a stalemate, due to the power of conservative political interests, according to Hughes (Hughes, 1993).

4.2.2 Deregulation and concentration of ownership

In their description of the saturation and stagnation of the electric power system in Sweden, Högselius/Kaijser focus on deregulation rather than on privatization. During this process, the state-owned power company Vattenfall also changed from a public agency to a transnational company (Högselius & Kaijser, 2007).

In Sweden, the ownership structure changed radically during the process toward deregulation. In 1991, there were eight large power companies and 256 other independent companies in the electric power business. In 2010, there were only three large power companies in Sweden: Vattenfall, E.ON, and Fortum. During the same period, large industries, such as SCA, Stora Enso, and ABB, sold out their power plants and left this branch. Of the 286 municipalities in Sweden, 207 had their own utility in 1991, private or owned by the municipality. Eleven years later, 100 of these municipalities had sold their power trading companies and 51 also their distribution grids (Högselius & Kaijser, 2007).

These books also describe the vertical integration and cooperation between companies that manufacture equipment for the power industry and the companies that operate it: AEG and Siemens in Germany, Westinghouse and General Electric in the United States, and Asea/ABB and its cooperating partner Vattenfall in Sweden.

4.2.3 Windpower ownership

Organizations linked to existing technologies will not initiate projects that are outside their discourse and not within their interest or perception, according to Henrik Lund (Lund, 2010). Power companies that are part of the DES will not develop windpower, and those who want to do that have thus to do it themselves, which makes them owners of these power plants.

The main conclusion from the comparative study on promotion in different EU countries (Lafferty & Ruud, 2008) is that the national contexts have a great impact. This is actually an important point also made by Hughes (Hughes, 1993). Such a comparison is interesting, as in, for example, Spain, the power companies linked to existing technologies actually play a dominant role also for the development of renewables (Dinica, 2008).

Ownership of wind turbines is described in some reports and papers, by Mark Bollinger about the United States (Bolinger, 2001), by Karl Sperling from Aalborg University about Denmark (Sperling, 2007), by Valentina Dinica about Spain (Dinica, 2008), and by several authors in Wind Energy International 2011/12 (World Wind Energy Association, 2011). Based on these and some other sources, a short overview of windpower ownership models in different countries is presented.

Denmark

In Denmark, farmers and cooperatives were the main customers of the windpower industry from the early 1980s and still are.

Until 2001, the electricity sector in Denmark was owned by consumer cooperatives and municipalities. It was regulated by a nonprofit legislation. Then the market was "liberalized," and two-thirds of the large power plants were sold to the state-owned power company DONG, and the remaining third to the state-owned Swedish power company Vattenfall. In 2005, the transmission system operator was bought by the state. The distribution system is, however, still owned by consumers and municipalities (Hvelplund, 2011).

To the end of the 1990s, cooperative windpower ownership was predominant. After that, the main new installed capacity has come from offshore windpower owned by large power companies.

To support local ownership, a new legislation was introduced in 2008, which gave the right of local and regional investors to attain 20 percent of onshore windpower plants at the cost price of the project. After 2008, several on- and offshore projects with cooperative and municipal ownership have been initiated.

Another interesting development is the increased collaboration between cooperatively owned windpower plants and CHP-plants to integrate these in a clever way into the power system (Hvelplund, 2011).

Germany

In Germany, a similar development with locally owned wind turbines started a decade later than in Denmark.

The German electricity market was liberalized in 1998. Four large companies—E.ON, RWE, Vattenfall Europe, and EnBW—own more than 80 percent of the power-generating capacity in Germany (Schroth, 2011). In Germany, about 51 percent of installed wind capacity is individually or cooperatively owned. Several hundred thousands of people have invested in citizens' wind farms across the country, and thousands of small and medium-size enterprises are running successful businesses in this new sector. Windpower has gained very high social acceptance in Germany, with the development of community wind farms playing a major role (Bayar, 2012).

Several large offshore windpower plants are under development, and these are owned by large power companies.

Sweden

In Sweden, a similar development with locally owned wind turbines started a decade later than in Denmark. From the early 1990s, single and small groups of wind turbines were developed and owned by cooperatives, farmers, and other private investors. Ownership categories and ownership models in Sweden are described in detail in Chapters 5 and 6.

The Netherlands

The Netherlands had around 2,400 MW windpower installed in 2012–2013.

The successful windpower development in Flevoland in the Netherlands, compared to other Dutch provinces (Bruinsma, 2008), also indicates that ownership is a crucial factor. In Flevoland, projects were initiated by farmers, and the regional authorities facilitated the process. This implies that farmers were owners of the wind turbines, at least in that part of the country. There are also some large offshore windpower plants outside the coast.

Spain

In Spain, public–private partnerships (i.e., joint investments by public agencies/municipalities and private companies) have played an important role during the first development phase in the 1990s. This is described in Section 4.2.6. In 2012, most windpower plants in Spain are owned by large power companies.

Portugal, France, Italy, and Greece

In 2012–2013, Portugal had installed as much as 4,500 MW windpower, which produce 17 percent of the electric power in the country. Development has been very fast; in 2004, there was not more than 500 MW installed.

Some of these windpower plants are very large, up to 240 MW. They are all operated by large power companies.

France, which is the most nuclear power-dependent country in the world, with 75 percent of the electric power from fifty-eight nuclear reactors, has more than 7,500 MW windpower installed in 2012–2013. Just like in Portugal and Italy, this development has taken place during the last ten years; in 2004 there was less than 400 MW installed in France. Large power companies dominate the market (World Wind Energy Association, 2011).

In Italy, with more than 8,000 MW installed in 2012–2013, private investors and power companies develop, own, and operate most windpower plants. The first player in the Italian windpower market is the international independent power-generating company International Power plc (IP), with interests in 33,000 MW of gross power production capacity located in twenty different countries. Other large developers and owners are Enel Green power, Edison Energie Speciali, and E.ON Italy. Ownership is thus dominated by power companies (Marchisio, 2011).

In Greece, windpower development has been rather slow, from 500 MW by the end of 2004 to 1,750 MW in 2012–2013. This growth has been attained by some major international players (e.g., Iberdrola, Enel, EdF, Endessa) in the local market but also major construction companies and private investors (World Wind Energy Association, 2011).

North America

Also in the United States, especially in the Midwest states of Minnesota, Wyoming, and so on, and in the province Ontario in Canada, many wind turbines are owned by farmers and cooperative associations (Martin, 2011). This is, however, just a few percent. In 2010, 83 percent of the installed windpower in the United States was owned and operated by independent power producers. Utilities only had 15 percent. The remaining 2 percent were made up of community wind ownership (Veilleux & Hamrick, 2011).

Most of the windpower plants in the United States are owned by large companies, which develop, build, and operate them to sell electricity to customers. NextEra Energy Resources is the largest, with more than 8,000 MW in operation. Many companies with the same strategy are from other countries such as Spain (Iberdrola, Acciona), France (EDF), and Germany (E.ON) (AWEA, 2013).

In Canada, the ownership structure is similar to the one in the United States, with Canadian and also some foreign windpower companies.

India

In India, most investors have been industrial companies that could benefit from the tax savings that the support systems offered. After the Electricity Act was passed in 2003, the electricity market has step by step been liberalized. In the last few years, international investors have been involved in developing very big windpower plants in India. Wind turbine manufacturers, such as Suzlon, also play an important role as project developers, investors, and operators of windpower plants in India (Bhambri, 2011).

China

In China, the electric power sector was liberalized in 2003. Major developers and investors are state-owned power companies or energy industry corporations that may be public or private, such as Longyan, Datang, Huaneng, and so on (Pengfei, 2011).

There are, however, also some foreign manufacturers that have built factories in China for the Chinese market and also for export to other countries.

4.2.4 *Summary of international comparison*

The ownership structure of windpower plants differ in different countries. In China with a very centralized planning economy, development is very fast and conducted by state-owned power companies but other actors are present as well, as a result of the liberalization and opening of the market. Some projects can be defined as public–private partnerships (PPPs).

In North America, the "build and operate" model is predominant, with many large foreign companies, some of which also manufacture wind turbines. These are vertically integrated companies.

In Spain, development started and took off owing to a PPP model, which also helped Spain to develop its own windpower industry. Some of these, such as Iberdrola and Acciona, have expanded to become leading actors on the world market.

In Denmark and Germany, local ownership and/or distributed ownership is a dominant model, so far. Also in Sweden, local ownership is rather common. Offshore windpower is in most countries developed by large power companies. These different ownership structures depend on historical factors as well as the energy policies in different countries.

4.2.5 *Community power*

Community power, a concept promoted by the World Wind Energy Association, can have several different meanings, depending on how the concept *community* is defined:

- local community: people living in same area;
- community of interest: people who share the same values;
- community of practioners: people working with same business; and
- community of consumers: electric power consumers.

All these four kinds of communities are relevant when it comes to windpower ownership models.

However, the concept *community power* has been defined by three points by the World Wind Energy Association Community Power workgroup (Martin, 2011).

A project can be defined as community power if at least two of these three criteria are fulfilled:

1 Local stakeholders own the majority or the entire project: a local individual or a group of local stakeholders, whether they are farmers, cooperatives, independent power producers, financial institutions, municipalities, schools, and the like, immediately or eventually the majority or all of a project;
2 voting control rests with the community-based organization: the community-based organization made up of local stakeholders with the majority of the voting rights concerning the decisions taken on the project; and
3 the majority of social and economic benefits are distributed locally.

The major part or all of the social and economic benefits are returned to the local community.

Based on experiences from the Middelgrunden offshore windpower plant in Copenhagen harbor, there are nine good reasons for local ownership, according to Puig I Boix: (Puig I Boix, 2011):

1 Local ownership results in more installed wind capacity.
2 Local ownership creates local dialogue and acceptance.
3 Local ownership raises public awareness.
4 Local ownership solves problems and conflicts.
5 Private investment promotes cheaper and better technologies.
6 Local production demands less transmission lines and saves electricity.
7 Local turbines are democratic.
8 Local production makes sustainable development understandable.
9 Local ownership gives people opportunity to act for sustainable development.

The goal of community power should be to provide a community with a decentralized source of renewable power that will embody the principles of sustainability and empower a community through substantial returns on investment that will go toward economic and social development in that community (Martin, 2011).

There are in fact many community power projects on line and operating in Denmark, Germany, Sweden, and some other countries. In Denmark, the government has promoted community power since the 1970s by different promotion systems and incentives. In 2011, more than 150,000 households in Danish communities own 86 percent of the total installed windpower capacity in Denmark (Wizelius, 2010/2012a). In 2003, some 90 percent of installed wind turbines in Germany were owned by citizens, representing more than 200,000 individuals (Martin, 2011).

Community ownership models have also been applied, on a lesser scale, in countries such as Canada, the United States, Japan, South Africa, and Australia (Martin, 2011).

To achieve sustainability of our energy systems, capacity needs to be built at all levels of society in an inclusionary manner. Clear goals, not only in terms of how much renewable energy we want and by what timeframe but also in terms of establishing mandates that define ownership structures that differ from those that are currently promoted by large players, are also needed.

By pairing sustainability evaluative criteria with ownership models and goals, a sustainable energy path can be assured, one that can alleviate society from threats of national and energy security economic instability and climate change. If fostered and implemented appropriately, community power can provide a useful mechanism that can ensure our future energy path is sustainable (Martin, 2011).

According to the World Wind Energy Association, community power has many advantages. In some countries, such as Denmark and Germany, such ownership models are common; in others, they hardly exist. One explanation for this may be the design of the promotion systems.

4.2.6 Promotion systems and ownership

Support systems used to promote renewables differ in different countries. In Denmark and Germany, they give the same economic benefit independent of ownership models. In Sweden, a recent change in the interpretation of the tax law can be claimed to discriminate cooperatives, as a "benefit tax" has to be paid on the difference between the electric power price from the cooperative and the market price (Skatteverket, 2008).

In India, the support system has been addressed by companies for so-called captive consumption. However, in December 2009, a generation-based incentive was introduced and, from March 2011, renewable energy certificates. This was supplemented with a new system based on production. The purposes of these reforms were to open up the market for international investors (Bhamri, 2011).

In the United States, the support system, with tax credits and accelerated depreciation, can be utilized only by companies with a balance sheet (Bolinger, 2011B). However, these limitations have been overcome by innovative organizational and financial solutions. In the United States, where the community wind sector is rather small, around 1 percent of installed capacity (Martin, 2011), these local windpower companies have used the so-called partnership flip structure to be able to use the U.S. support system (Bolinger, 2011a).

This means that the local group of investors finds a partner that can utilize the tax benefits and accelerated depreciation to co-own the project. This partner, the tax equity investor, provides most of the equity for the project and will during the first years of operation get most of the cash and tax benefits from the windpower plant. When most of the tax benefits have been utilized and the tax equity investor has gotten the returns on the investment agreed on in advance, the allocation of cash and benefits "flip over" to the local owners. In most cases, the local owners then have the option to buy out the tax equity investors, whose interest in the project has diminished (Bolinger, 2011a).

This flip-model was developed by community power projects in Minnesota and has been used since the year 2000. From 2010, some community wind projects have found new creative ways to finance their projects, such as a combination of low-cost government debt and local tax equity or a sale/leaseback structure and other new solutions. These have been facilitated by recent policy changes, which not have been specifically addressed to the wind sector.

Most important is the American Recovery and Reinvestment Act from 2009, which enables projects to choose either a 30-percent investment tax credit or a 30-percent cash grant instead of the federal incentive that has been available to windpower projects in the United States, a ten-year production tax credit. This flexibility makes it possible for windpower projects to use leasing models for financing. These new financing models are described in detail in articles by Mark Bolinger (Bolinger, 2011b).

For community wind projects seeking financing in the United States, necessity has been the mother of innovation, Mark Bollinger concludes. This innovation of financing models has also been apparent in Sweden. To be able to sell the wind turbines that project developers have gotten permissions for, the developing companies have first found a way to apply the cooperative ownership model to wind turbines, developed this model over the years, and also offered several other investment schemes to farmers, enterprises, and local inhabitants. These are described in Chapter 5, Ownership Models in Sweden.

4.2.7 Public–private partnerships in Spain

Spain has had an impressive development of windpower from a rather modest 442 MW in 1997 to 8,511 MW in 2004, 15,145 MW in 2007, and as much as 22,796 MW in 2012 (EWEA, 2013). That year, Spain was number four in the world when it came to installed capacity, number three in respect of windpower penetration, and number two in windpower per capita. Just as for Denmark, windpower has been a success story for Spain.

This success has by most researchers been credited to the efficient feed-in-tariff system that has been used (Ackermann, 2001; Gan, Eskeland, & Kolshus, 2007). Valentina Dinica, from University of Twente in the Netherlands, argues convincingly against this simple thesis and claims that the source of this success has been the ownership models used in Spain (Dinica, 2008).

Public–private partnerships have played a crucial role for windpower development in Spain, from the early 1990s and until the take-off that took Spain to one of the world leaders in the windpower business, not only in installed capacity but also as a manufacturer and worldwide wind developer.

This narrow policy descriptive-research style, using econometric models, has ignored other non-economic policies and strategies, Dinica argues. Renewable energy feed-in tariffs have no doubt been an efficient tool to promote investments in renewable energy sources, creating investment security. However, in Spain, such a system, with investment security, was not introduced until 2004. Dinica points out that the Spanish feed-in tariff system had up to that time a very unattractive design with high economic risks: only five-year purchase contracts, far from the lifetime of a wind turbine.

There are many other contextual factors, such as grassroots movements, institutional contexts, spatial planning policies, and so on, which can play an important role for the diffusion of renewable energy. Dinica argues that the success story of windpower in Spain can be explained by

> ...a policy—implemented on all governmental levels—of stimulating investments by means of public-private partnerships (PPPs).
>
> (Dinica, 2008)

In Spain, there are three levels of government: national, regional, and local. PPPs are legal partnerships between public actors (agencies or departments of local, regional, or national governments) and private actors, for joint equity investments in windpower projects. By sharing investments with public actors, private investors perceived a reduced risk for their investments in windpower projects (Dinica, 2008).

Investments based on partnerships during the period 1980–2000 were responsible for 95.7 percent of the close to 3,000 MW windpower that was put into operation during that period. All of these were, however, not public–private partnerships; some were partnerships between two or more private companies (Dinica, 2003).

The national Institute for Energy Saving and Diversification (IDEA), established in 1984, played a key role for the development of the public–private partnership model. One of its tasks was to draft, implement, and monitor the national renewable energy plans in Spain. Up to the mid-1990s, IDEA was the main public element in PPPs, investing in windpower projects (Dinica, 2008).

The purpose as well as the effects of a public agency's investments in windpower projects was to raise the interest of private companies to do the same. The private partners in these PPPs were power companies, industrial corporations, banks, insurance corporations, and institutional investors. IDEA played the role as diffusion catalyst for windpower. It was not the financial contribution but the trust in legal support and technology that reduced risk that was important for the private partners to join PPPs.

During the early commercialization phase during the 1990s, project-vehicle partnership (i.e., companies were formed for one specific windpower project) were dominating and typically were formed by IDEA offering political support, a wind turbine manufacturing company supplying technology and guaranteeing quality, a power company guaranteeing grid connection, and a regional or local authority securing permits and social approvals and that also could be the land owner.

Together, these PPPs managed in close dialogue with one another and the government to form a support system securing a cost-efficient windpower development with low economic risk. From 2000 and on, PPPs developed into wind-specialized partnerships (working with windpower project portfolios) or into partnerships diversifying to other renewables (Dinica, 2008).

Another very interesting spin-off of PPPs in Spain has been the technical integration of windpower plants into the Spanish electric power system. To enable a large penetration of windpower in the system, technical regulations had to be added to the support system, for productions forecasts, fault ride through capability, voltage control, and power regulation.

Such demands from grid operators have in most countries been considered as "non-technical barriers" to windpower development (Greenpeace, 2005). In the Spanish PPPs, all involved actors (manufacturers, grid operators, etc.) have a stake in the successful integration of windpower in the system. As they jointly own the windpower plants, there was a consensus on these demands, and they have also been implemented (Abbad, 2010).

4.2.8 New perspectives on ownership in Denmark

In the report *Large-scale Windpower in Denmark—New Perspectives on Ownership* (Sperling, 2007), Karl Sperling from Aalborg University discusses different ownership models used in Denmark. He evaluates ownership models in relation to *local acceptance* of windpower projects.

Sperling presents a conceptual model that describes factors that influence local acceptance. Ownership is identified as one significant factor in this model.

He identifies five different ownership models used in Denmark:

- individual;
- cooperative;
- power company;
- local distribution company; and
- municipality.

Sperling then sets context-dependent criteria for new forms of ownership and combines these into four new models of *co-ownership*:

- inland model;
- offshore model;
- coastal-offshore model; and
- island model.

His idea is that there are benefits when several different actors own a windpower plant together.

The inland model

Windpower plants sited on the mainland—Jutland and the larger islands of Denmark—should be co-owned by a cooperative and a municipality and in some cases also by the local distribution company. Sperling makes the assumption that careful siting and local acceptance will be the main requirements for successful implementation of windpower at inland locations. By combining cooperative and municipal ownership, it will be possible to find suitable sites and get maximum local support for projects.

The offshore model

Windpower plants far from the shore should be owned by power companies and individual investors. Owing to high costs, investment security is the main challenge. Local communities are far away, so local acceptance will not be a problem. Power companies have the financial and technical capacity to develop such projects. These are national macro-level projects, and the national public may be involved by offering citizens to invest in shares.

The coastal-offshore model

Offshore windpower plants close to shore are an intermediate model between offshore and inland models. As these wind turbines will be quite visible, local and maybe also regional acceptance is necessary. These should be co-owned by a cooperative, with shareholders from the region to gain acceptance, and by the local distribution company and eventually also by a power company, owing to rather high investment costs.

The island model

One island is often equal to one local community. Windpower plants on islands should be co-owned by the municipality and local cooperatives and eventually also by individual investors. On islands with unique landscapes, local acceptance is crucial. This can be gained by cooperative ownership. Regional development is important for all islands, so the municipality should be owner as well. If the inhabitants of an island cannot cover all of the investment, the cooperative can offer membership to a wider public, or shares could be sold to individuals.

There are case studies of the coastal offshore model, Middelgrunden in Copenhagen harbor, and of the island model, Samsö, where also a coastal-offshore wind farm is in operation.

The conclusions Sperling draws from the case studies is that local anchorage of the responsible planning authority may be crucial to gain local acceptance and overall project permission in these forms of co-ownership. This local character of planning is closely associated with local forms of ownership.

Sperling's report is closely related to the topic of this thesis, although Sperling discusses ownership in relation to local acceptance, while I discuss ownership in relation to local and economic benefits.

4.2.9 Ownership and acceptance (support)

Ownership is identified by Sperling as a major factor influencing the local acceptance of windpower projects in a community. His aim is to find forms of ownership that are likely to attract local acceptance and how existing forms of ownership can be combined to ensure this and secure development of windpower projects (Sperling, 2007). This is based on the assumptions that existing forms of ownership cannot fulfill the requirements of the *second phase* of Danish windpower development.

Sperling also points out that the term *acceptance* in a top-down fashion assigns a somewhat passive role to the (local) actors who are "affected" by a specific windpower development. He suggests that the term *support* should be used instead. Support is associated with active and positive (bottom-up) attitudes. (He keeps the term *acceptance*, as it is more common in the literature.) I fully agree with this point and will use *support* as a synonym to *acceptance* in my own writing.

The Dutch researcher Maarten Wolsink emphasizes the planning regime and the degree of local ownership as variables affecting positive windpower project decisions (Wolsink, 2007a, 2007b).

Attitudes to windpower projects can also be motivated by *self-interest*. Economic benefits that increase personal benefits can change attitudes (Bell, 2005; Wolsink, 2006). Shareholding in community-owned or privately-owned wind energy developments could be a response to the "self-interest", according to the researchers Bell, Gray, and Haggett (Bell, 2005).

According to Sperling, there seems to be a direct relation between ownership type and degree of local support for windpower developments. This is supported by Frede Hvelpelund, who states that acceptance of wind turbines will depend on the extent to which some of the economic benefits of the development are given to the people who have to live with the impacts of these turbines (Hvelpelund, 2006).

The Japanese researchers Maruyama, Nishiko, and Iida show that environmental, economic, and social commitment may all be reasons for people to invest in windpower and that community projects make it possible to combine these in a single project (Maruyama, Nishikido, & Iida, 2007).

There are many reasons for people in local communities in different parts of the world to support windpower and many reasons for them to oppose windpower projects. Ownership seems to be an important aspect in this regard.

4.2.10 Fairness and ownership

Support of and opposition to windpower are mostly described at the micro-level. According to Maarten Wolsink, issues of fairness and equity significantly influence attitudes rather than NIMBY (not in my backyard) motives (Wolsink, 2007).

Elinor Ostrom also identifies fairness and trust as important variables to create robust systems for management of common pool resources (Ostrom, 2005).

There may be an opinion on the lack of fairness and equity on the macro-level as well concerning, for example, the electric power market. Although the political parties in parliament do not question the fairness of the electric power market, many other significant actors do: researchers, economy journalists, ordinary household customers, industrial corporations, and even different state agencies (see Sections 3.4.7–3.4.10).

In Sweden, this perceived or real lack of fairness of the electric power market is also a motivation for different actors to invest in their own power plants. The significance of this aspect will be one of the questions of the investigation of windpower ownership models.

5 Ownership models in Sweden

The technical development of windpower has been very fast, and the size of wind turbines has increased. The investment cost for a wind turbine has also increased dramatically, from 1 to 2 million SEK for a 150-kW turbine in the early 1990s to 30 to 35 million SEK for a 2-MW turbine in 2012. At most sites, it makes sense to install a group of turbines rather than just a single one, which makes the investment costs rise to 100 to 500 million SEK for a typical site in Sweden in 2012.

Projects of this size may seem too big to be handled by local cooperative associations or local limited companies. However, these owner categories are still growing, which is shown in Chapter 6.

The answer to the first research questions—*Which ownership models for windpower are used in Sweden, and how are they designed?*—can explain how small investors can manage to be partners also in very large windpower projects.

5.1 Windpower business structure

Before coming to the research question, the ownership models have to be put into a context. The development process from idea to an operating windpower plant passes through several stages with different actors involved (see Figure 5.1). Interesting questions are at what stage and by which actor these ownership models are decided, and how and by which actor they have been developed.

Windpower project developers can apply several different business models. A company in the windpower business can specialize in one phase of development, while the other may work through the entire chain, from the feasibility study to the operation of windpower plants, and take the earnings from sale of electric power.

There are significant differences in the company structure in different countries. In Europe, there are many companies that develop projects and make an exit either when the permissions have been granted or when turn-key plants are installed and online. In India, manufacturers often develop their

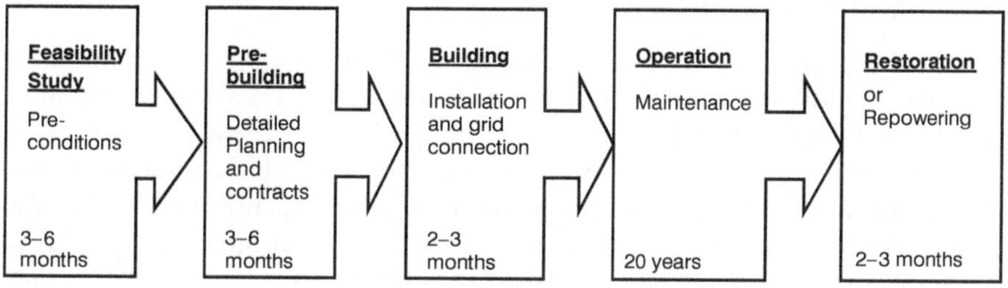

Figure 5.1 Development stages of a windpower project (illustration from Wizelius, 2012).

own projects, sell them when online, but also take charge of the operation during the lifetime of windpower plants. Many large power companies, such as Iberdrola from Spain, Dong from Denmark, and Vattenfall from Sweden, develop large windpower plants on their own for operation and sale of power.

The highest risk is usually in the first stages of development, when there is uncertainty whether the project will get the necessary permissions and whether the windpower plant can be connected to the grid. Companies specialized in project development start with land lease contracts to secure access to land suitable for windpower plants. Some will even sell these contracts to other developers and make a very early exit. Most will, however, make a project description and an environment impact assessment and then apply for permission to build a windpower plant.

When permissions have been granted, there is an exit point; the project developer may sell the project to another company. To enter the building phase, there are other competencies needed and also a lot more capital and financing skills. This is, thus, also the entry point for companies who are specialized in building windpower plants.

Finally, there are consultants who will do this work for a customer, often large power companies, but sometimes also on behalf of land owners who do not have these in-house competencies (Wizelius, 2012).

5.2 Development of ownership models

Before the different ownership models are described, I will give a short summary of how and why they have been developed. Different models have been developed by project developers, with the purpose to finance their windpower projects.

To get the facts, I had interviews with two persons who have been engaged in windpower development in Sweden from the start in the early 1990s: Andreas Wickman, CEO of Wickman Wind AB, and Thomas Sirland, CEO of Siral Co AB.

5.2.1 Real estate communes

Andreas Wickman was one of the founders of the windpower development company Vindkompaniet. This company found that the real estate commune would be a good ownership model for windpower. A real estate commune is based on common law and communal ownership of physical resources in rural areas, such as fishing harbors, private roads, gravel pits, and so on, and is also used for parking lots and playing grounds. These are common properties of a group of real estate owners. The shares in the real estate commune become parts of the real estates, not of the owners personally (Österberg, 2013).

In the early 1990s, no banks would finance wind turbines. Using the real estate commune, the share owners could use their land, farms, or houses as mortgage for loans. Using this model for windpower, the cooperative's entire investment was paid up front. This was the only way to finance windpower projects in the early 1990s, according to Andreas Wickman.

After a few years, and when the market for wind cooperatives on Gotland was saturated, Vindkompaniet started to develop turn-key windpower plants that were sold to companies, farmers, and private investors. During the last few years, Andreas Wickman has organized some large repowering projects at Näsudden on Gotland and developed a new ownership model, the leasing model. This will be described in Section 5.8.5 of this chapter.

5.2.2 Partnership contracts

Thomas Sirland is one of the owners of the windpower project development company Siral Co AB on Gotland. He was also one of the founders of the windpower limited company Slitevind AB. This company was formed in September 1992 to be able to finance a group of three wind turbines on Gotland. Investors bought shares in the company, and this way the money for the investment was raised. For each new project, new issues of shares were released, Thomas Sirland explains.

The company Siral has also during the years developed its own ownership model based on a *partnership contract* between investors. This group is formed when a project has the necessary permissions and it is time to start with procurements. This ownership model, the partnership contract model, is described in Section 5.8.4 of this chapter.

5.3 Corporate business models

Power companies and windpower limited companies have a quite regular business model. The revenues from sale of power and from green certificates are used to cover capital costs and the cost of operation and maintenance, hopefully leaving a reasonable profit.

There are basically two ways of financing a windpower project. One is corporate financing, which can be used by large companies who have the capital needed for the investment available within the company. The wind

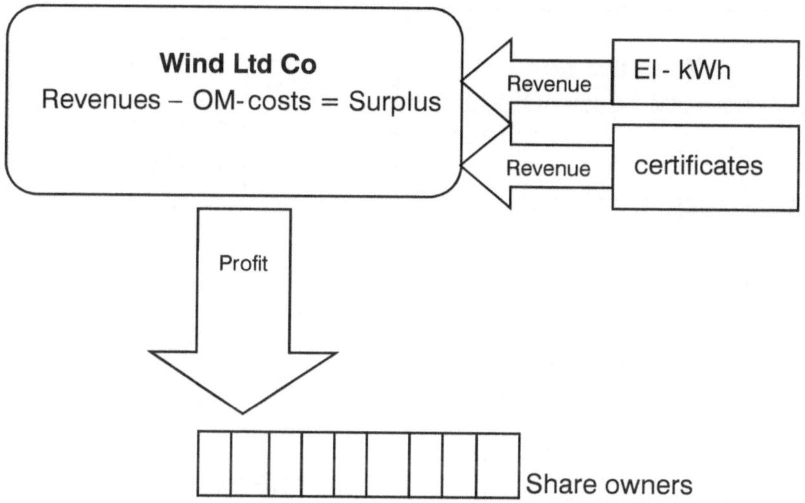

Figure 5.2 Corporate business model with 100 percent financing. All electric power from the wind turbine and the certificates are sold to the market. Surplus is reinvested or paid as dividends to shareholders

turbines will then just be new machines used by the company, and the costs and revenues will be integrated in the company balance sheet.

The other way is called *project financing*, which can be used by large companies as well. In this case, the windpower project is treated like an independent economic entity (Redlinger, Andersen, & Morthorst, 2002).

In this chapter, only models for project financing will be described.

5.3.1 Windpower limited company with 100 percent equity

When all capital for the investment is raised from the sale of shares, the model is quite simple (see Figure 5.2).

5.3.2 Windpower limited company with equity and debt

If banks and other financial institutions can admit loans on reasonable terms, windpower projects can be financed by a combination of equity and debt, which makes it possible to develop a larger windpower plant (see Figure 5.3).

The terms for loans depend on the calculated revenues and risks. The higher the risk, the higher the interest rate, and lower share of the total investment will be financed by loans.

5.3.3 Windpower limited company with equity and loans

If it is difficult to get loans from banks or other financial institutions or the terms offered are unfavorable, a windpower limited company can add

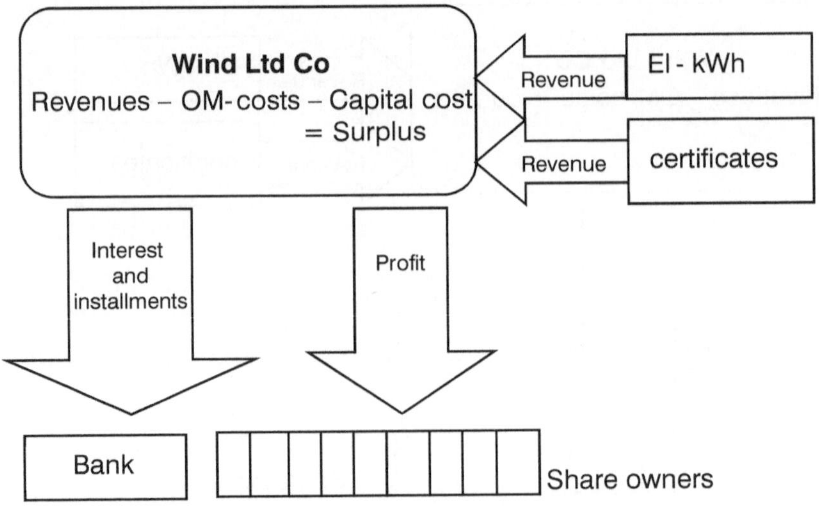

Figure 5.3 Corporate business model with equity/debt financing. Some of the investment is financed by loans from banks or other financial institutions. All electric power from the wind turbine and the certificates are sold to the market. Surplus is reinvested or paid as dividends to shareholders

a demand on shareholders to offer a certain amount as a loan for each share, to be allowed to buy shares. This loan from shareholders is then paid back according to a plan, usually within a few years. This model was quite common in the 1990s when it was difficult to get financing from banks, but it is still used by some companies (see Figure 5.4).

5.4 Cooperative business models

Windpower cooperatives in Sweden are organized in a number of different ways. The first cooperatives in Sweden, formed on Gotland, used the real estate commune (samfällighet) as organization model.

These associations, traditionally used for private roads, local harbors, watersheds, and so on, are open only to real estate owners, since the common asset (road, harbor, wind turbine) is tied to the estate and not to a person. As this model is no longer used, it is not described further in this chapter.

When Göteborg Energi (the municipal utility in Gothenburg) was planning to start a windpower cooperative for their customers (i.e., household consumers in Gothenburg), a thorough investigation was made to find an organizational form open to all. They found that a cooperative economic association would be the best option.

The windpower cooperative Göteborgsvind ek för was formed in 1992, with two wind turbines in Gothenburg harbor. Two more wind turbines were added in 1993 and 1996, and the cooperative association has 1,100 members. This is the only type of association used for windpower cooperatives today.

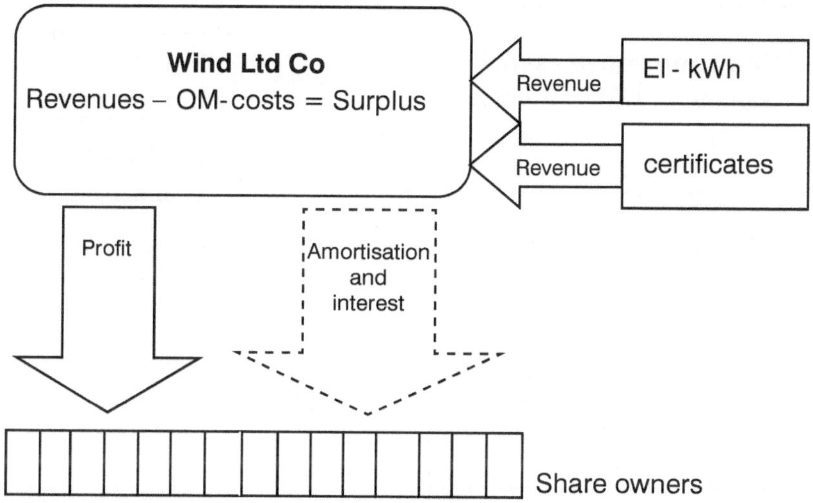

Figure 5.4 Corporate business model with equity investor financing. Some of the investment is financed by loans from shareholders. These may get better terms than the company for loans, using real estate (farms, land, etc.) as deposits. All electric power from the wind turbine and the certificates are sold to the market. The loans are paid back according to plan, with interest. Surplus is reinvested or paid as dividends to shareholders

These economic associations have, however, used three different models:

- production to cover members' consumption with annual refunds;
- production to cover members' consumption by net accounting; and
- production for selling power (business venture).

The first two of these three models are unique for Sweden but could be used also in other countries.

5.4.1 Cooperatives with annual repayment

This is the classical model used by real estate communes as well as cooperative associations in the early 1990s. Power that is produced by the members' own power plant and all used by the members for their own consumption should not be considered as a business venture and, therefore, the cooperative should not have to pay any company tax and the members not pay any energy tax. The condition for this is, of course, that no member owns shares corresponding to more than his own consumption. If just one member will do that, the entire co-op will be considered to do business and thus have to pay these taxes.

Therefore, members were not allowed to own more than what corresponds to about 80 percent of their expected power consumption. That conservative figure was used since the annual power production will vary within around 20 percent in different years.

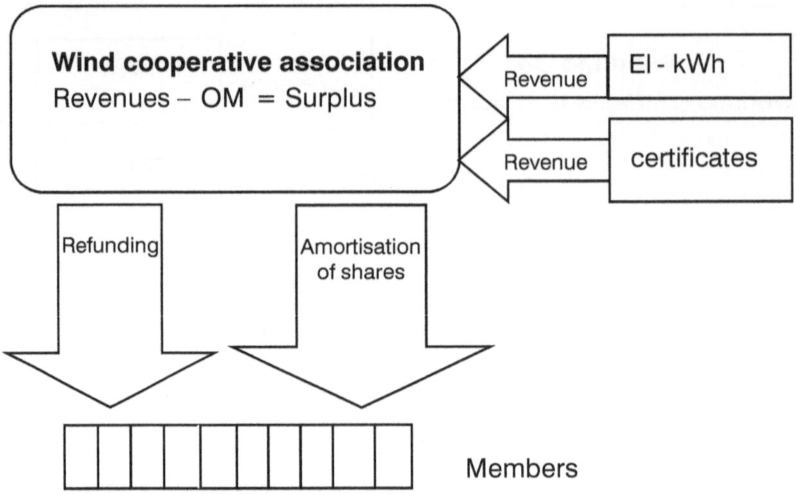

Figure 5.5 The classic coop model. This model is just like a producer's cooperative. If the members' share of the production is less than their own consumption, there is no company tax. All electric power from the wind turbines and the certificates are sold to the market. Members pay their normal electric power bills, and once a year members get refunds from the wind cooperative

Before the deregulation of the electric power market, all power had to be sold to the power company or utility that had concession to distribute and trade electric power in the area, at fixed price. After the deregulation, the same models were kept, but they could make PPAs with any power trading company in Sweden.

The revenues, after the certificate system was introduced in 2003, consist of incomes from the sale of electric power and certificates. Each year, members get refunds from the cooperative in relation to their shares. Most cooperatives using this model also pay back the initial investment according to a depreciation plan set for twenty years, which is the estimated technical lifetime of a wind turbine (see Figure 5.5).

Unfortunately, not all tax authorities accept the idea that the power is for the members' own consumption. In some counties, tax agencies consider this as a business model and demand that the cooperative association pay company tax on the annual surplus. When cooperatives have had this tried in courts, this has been confirmed by judgments. In other counties, tax agencies accept this as a "fictive transmission" and do not charge any company tax on annual profits.

The value-added tax (VAT) is handled by the cooperative association. The VAT on the investment is deducted, and the VAT from sale of power is accounted for by the cooperative, not by the members.

5.4.2 Cooperatives with net accounting

This model was developed by the windpower cooperative Kvarkenvinden ekonomisk förening and the municipal utility Umeå Energi (Wizelius, 2010/2012a).

Net accounting means that the electric power delivered by the wind turbines of the windpower cooperative's member is deducted from the member's power bill from the utility, so that the member/customer will pay only for the remaining power delivered by the utility to fill the gap between the customer's/member's total power consumption and the power from the windpower plant.

Windpower cooperative shares usually correspond to 1,000 kWh/year, according to the calculated production. Members are recommended not to buy more than around 80 percent of their own power consumption.[1] If a member's shares during a good wind year produce more than the member's consumption, the surplus is lost (or rather is not paid for and is given to the utility).

The idea is that the grid operator transmits the power from the windpower plant to the members/customers, so this wind-generated power is neither sold nor bought by anyone. It is just transmitted. The only cost is the grid fee. The balance power (i.e., power used when the wind turbines produce less than a member/customer consumes) is delivered and sold by the utility.

This model is based on a close cooperation and an agreement between the cooperative association and the power trading company. All members have to be customers of this power trading company.

The cooperative pays a fee to the power trading company for the service of balance responsibility, a few öre per kWh. The cooperative sells the certificates, which brings some revenue. It also has to cover all costs for operation and maintenance of the windpower plant, such as service, maintenance, insurances, land lease fees, administration, and so on. This is the cost price for the wind-generated electric power, and this cost price is decided each year at the association's annual member meeting (see Figure 5.6).

Certificate prices are very volatile. During years with high certificate prices, the member fee can be close to zero, or even negative. However, members also have a private capital cost, from the investment made in the shares, so there is still a cost for the electric power from the wind. Members who leave the association will get back the money paid for the shares but only its nominal value, without any interest.

A controversial question is how to calculate the balance. From a bookkeeping point of view, there is no problem to use a full-year balance. However, as this gives an economic advantage, tax authorities may have stricter demands. Formally, it should be a real-time balance. So practice varies; some power traders apply a monthly balance; Umeå Energy applies a three-month balance for Kvarkenvinden; and others even an hourly balance.

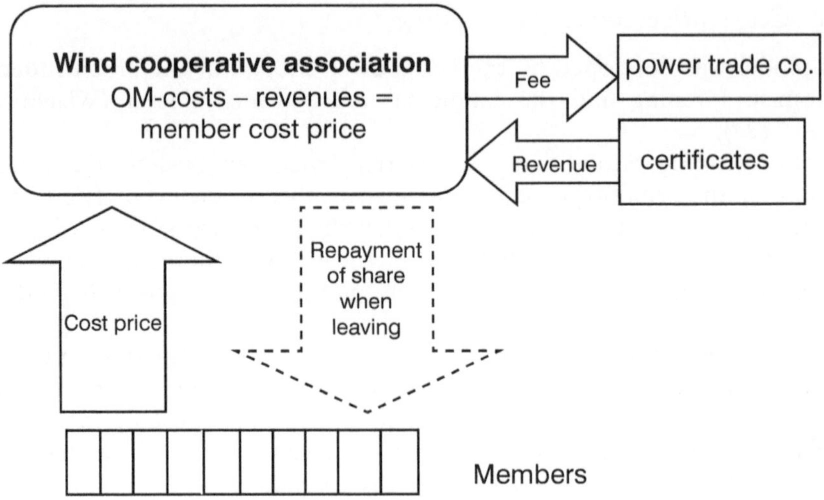

Figure 5.6 Net accounting/transmission. With this model, the power (kWh) delivered by the wind turbine is deducted by the power trading company from the members' power bills. All members thus have to be customers of the same power trading company. This company also delivers the power for balance, when the wind turbines produce less than the members' consumption. If there is a surplus, this will not be paid for or saved for later use. Members are recommended to buy a maximum share, which corresponds to around 80 percent to 90 percent of their own estimated power consumption

The shorter the balance used, the lower share of the power consumption can be used by customers. For wind cooperatives, however, either a monthly, quarterly, or annual balance is usual.

Another issue is the energy tax. According to the electricity law, power produced by an end user does not pay energy tax. This does, however, not apply to members of cooperatives, according to a verdict in the tax court. Rules are, however, very favorable anyway. It has so far been cheaper to get power from a cooperative with net accounting than from the market.

5.4.3 *Producer cooperatives*

A producer cooperative works just like a wind limited company. It is a strict business venture where all power and certificates are sold on the market, and the surplus is either reinvested in new wind turbines or refunded to the members after the association has paid company tax. The members have to pay tax on the annual revenues as income of capital (see Figure 5.7).

5.4.4 *National cooperatives*

Another innovation has been the formation of *national* cooperatives. There are three companies using this concept: Sveriges Vindkraftskooperativ ek förening, O2 Energi, and Telge Energi. Most windpower cooperatives can

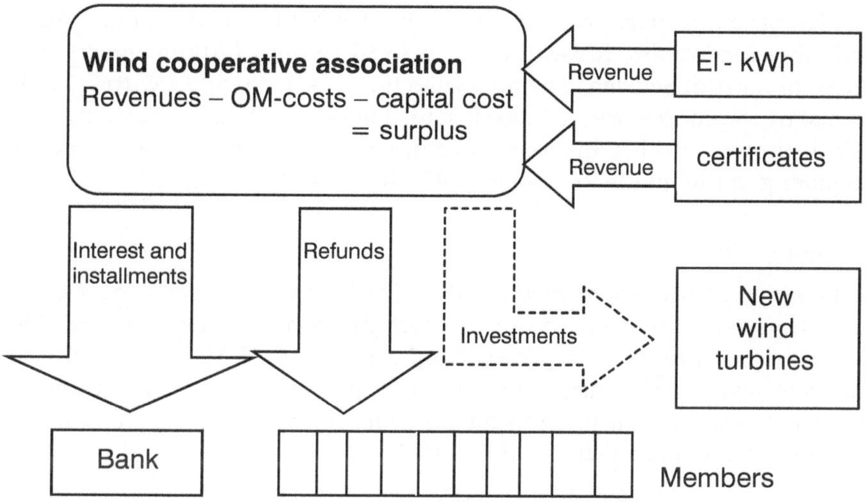

Figure 5.7 Producer cooperatives. A producer cooperative works just like a limited company. All electric power and certificates are sold on the market. The annual surplus (profit) is either returned to members as dividends or reinvested in more wind turbines

actually have members from all over Sweden, unless there are rules about this in the statutes.

Before deregulation of the electric power market, this was not possible. For real estate communes, there were rules that restricted membership geographically to the same Land Survey district where the wind turbine was installed. Other windpower cooperatives (i.e., economic associations) have to be open to all to become members. Most windpower cooperatives, then as well as now, market their shares locally or regionally. These three companies, however, consider all of Sweden as their market and have their windpower plants spread all over Sweden as well.

In these cooperatives, new members have to buy shares corresponding to 1,000 kWh/year, and the current price is around 5,500 to 7,500 SEK ($700–$900). Members will pay the production cost for the power they use, far below the market price (there is, however, a capital cost for the shares). A member who wants to leave the cooperative gets back the money paid for the shares. This is the same principle as for all cooperatives; however, in the national cooperatives, members get shares in a "portfolio" of windpower plants, which is not restricted to the technical lifetime of specific wind turbines.

Sveriges vindkraftskooperativ ek för

Sveriges vindkraftskooperativ ek för (SVEF; Swedish Windpower Cooperative Association) was initiated by the municipal utility in Falkenberg that has been a pioneer not only for windpower but for solar heating systems and other renewables. It was formed in 1998 and has 1,800 members from different parts of Sweden. The nine wind turbines, from 600 kW to 2.1 MW, owned by the cooperative are also installed in different parts of Sweden: one in Falkenberg, another in the far north, still another on Gotland, and the remaining in the middle of Sweden (Wizelius, 2010/2012a).

O2 Energi AB

A similar company was formed in 2006, O2 Energi AB, based in Stockholm and affiliated to the windpower project developer Vindkompaniet. This company has formed a separate cooperative with 3,500 members with shares in seven 2-MW turbines in different parts of Sweden. The company has many new wind turbines in the pipeline, so no one has to stand in line and wait to become a member (O2, 2012).

Telge Energi AB

Telge Energi is a municipal utility in Södertälje, 35 km south of Stockholm. After deregulation of the electric power market, the utility has worked as a power trading company with customers from all parts of Sweden. In 2009, the company was the first in Sweden to sell 100 percent clean electric power only from windpower plants.

In 2010, Telge Energi started Telge Energi Vind ek för, a windpower cooperative, and offered their customers to buy shares and become members. It now owns two large wind turbines in different parts of Sweden.

Most wind cooperatives offer shares for a minimum of 1,000 kWh/year. Telge Energi has divided the shares into smaller blocks, so it is possible to buy shares of only 100 kWh/year. For urban households with low electric power consumption, this makes it possible to adapt the shares better to the actual power consumption (Sundelius, 2013).

5.4.5 Cooperative members

In the early 1990s, all members/part-owners in windpower cooperatives were private persons: households. However, according to the law on economic associations, legal persons (i.e., companies) can also be members of economic associations. I have not analyzed the membership lists of all windpower cooperatives in Sweden, which are public, so it could be done. I have, however, asked cooperative administrators during an inquiry made in 2008, while writing the manual for windpower cooperatives (Wizelius, 2010/2012a), which kind of members the cooperatives had.

Based on this, I can state that the share of legal persons (companies, etc.) has increased and is now a dominant subcategory among members/owners of windpower cooperatives. There are many food-shop owners from the retail chain ICA, other SMEs, municipal housing companies, and private housing societies. Some cooperatives, such as the national cooperative O2 Energi, are marketing shares specifically to such companies and associations.

5.4.6 Municipal promotion of cooperatives

Many municipal utilities have initiated and cooperated with cooperative associations that own and operate windpower plants. Falkenberg Energi has already been mentioned. Umeå Energi has had a close cooperation with the cooperative Kvarkenvind ek för, founded in 1998, which has expanded rapidly in the last years, adding several multi-MW turbines, and has, beside private consumers, municipalities as members and small businesses (Wizelius, 2010/2012b).

Göteborg Energi created the cooperative Göteborgsvind ek för already in 1992. Kalmar Energi established the cooperative Kalmarsundvind ek för and, in 2009, it also won a tender to administrate the cooperative Kyrkvinden, which owns one of the turbines in the offshore wind farm in the lake Vänern (Larsson, 2009).

5.5 Real estate companies

New actors on the windpower market are private as well as municipal real estate companies. This all started in 2006, when the municipal housing company in Askersund, after a thorough investigation, bought its own wind turbine. The idea behind this was for the company to produce its own electric power. If power is produced, transmitted, and finally used but never bought or sold on the way, there is no electricity tax.

This was a very good business idea that reduced the cost for energy significantly. Many other municipal real estate companies have followed this example; one of the latest is Sigtunahem, which now owns a wind turbine in Skåne (Sigtunahem, 2013), and the municipality owns another two wind turbines to cover its electric power consumption at a competitive price (Sigtuna, 2012).

This model used by municipal companies is the same as the net accounting model (see Figure 5.6). The difference is that wind cooperatives have to pay energy tax, while companies do not, according to a verdict by the Swedish tax authorities. To use this model with tax exemption, the wind turbine has to be owned by the same corporate body that will use the power. It has to use corporate financing not project financing (see Figure 5.8).

It is also somewhat unclear how to manage the balance. All power from the wind turbine(s) is fed into the power grid, where it is mixed with all other power. How much is fed into the grid is, however, measured, down to one hour. A power consumer, in the case buildings where the power will be used, also has a meter where the grid operator measures the consumption that is registered down to one hour. The spot price is also set for each hour. According to tax authorities, the demand for balance will not be less than

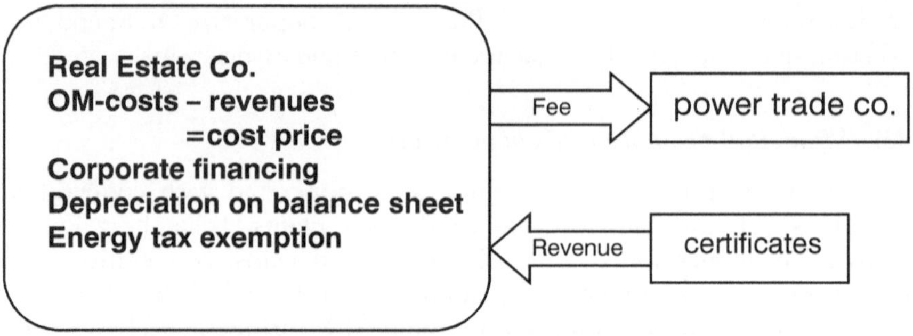

Figure 5.8 Real estate company with windpower for its own consumption. As long as no power is bought or sold, a company can transmit power on the grid from its own power plant and use the power without having to pay any energy tax. However, production and consumption have to be in balance (simultaneous)

one hour. It did not say whether it could be acceptable with a daily, monthly, or annual balance.

The private real estate company Wallenstam, based in Gothenburg, has several wind turbines on line to supply their houses as well as tenants with renewable wind-generated electric power. The company gets all the energy for its houses and buildings from renewables (Wallenstam, 2014).

Although it is possible also for private companies to use the model described above, Wallenstam has used another model. It has created the daughter company Wallenstam Naturenergi AB, which is a power trading company. The windpower plants developed by the company, Wallenstam Skogaryd AB, Wallenstam Skuggetorp AB, and so on, are organized as daughter companies to Wallenstam Naturenergi AB (see Figure 5.9).

Wallenstam buys all power from the windpower plants at spot price and then sells it, also at spot price, to Wallenstam AB to cover the power consumption in their rental properties. The cost of electricity is included in the rent of tenants, so the interface between grid and consumer is not at the single apartments but at the buildings.[2] The addition to the rent from electric power is set in relation to the tenants' power consumption.

5.6 Industrial companies

Industrial users who feed the power from the wind turbines directly into their factories will get the power at cost price. Neither energy tax nor any grid fees have to be paid.

Industrial corporations, who operate grid-connected wind farms, usually sell the power on the market. As the price for the power that is sold varies the same way as the power that is purchased on the market, this arrangement works as a "hedge."

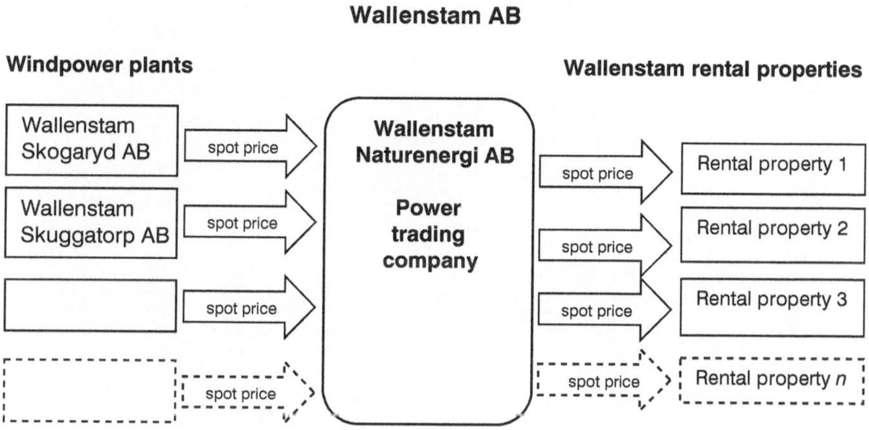

Figure 5.9 Real estate company with windpower plants and trading. The mother company, Wallenstam AB, is the owner of the power trading company Wallenstam Naturenergi AB and of all the rental properties. The windpower plants are daughter companies of Wallenstam Naturenergi AB

5.7 Mixed ownership

Many large windpower plants as well as large (2- to 3-MW) turbines have a mixed ownership; 50 percent can be owned by a cooperative, 25 percent by a local farmer, and the rest by the utility. Any mix can do.

Project developers have been very creative in finding good investment models that enable local and/or mixed ownership. In wind farms with many wind turbines on the same site, the single turbines can have different owners; some owned by power companies, others by cooperatives, by local companies, and so on. One example of mixed ownership is the wind farm Högberget in the county Dalarna (see Figure 5.10).

Not only wind farms but also single turbines can have a mixed ownership. On Gotland, for example, a group of eight farmers share a turbine named Group 8. Each of them has a company for their farms, and the (part of) wind turbine is treated in the farms' bookkeeping as a machine inventory. These eight companies have formed a firm that takes care of the administration of the turbine (i.e., pays bills and distributes the annual revenue to the eight owners). The paperwork can be handled by an external bookkeeper.

5.8 Umbrella companies

In a windpower plant, turbines will affect one another, owing to the wind wakes. Some will have a better position and a higher production than others. This would cause disputes when the turbines in the windpower plants are sold.

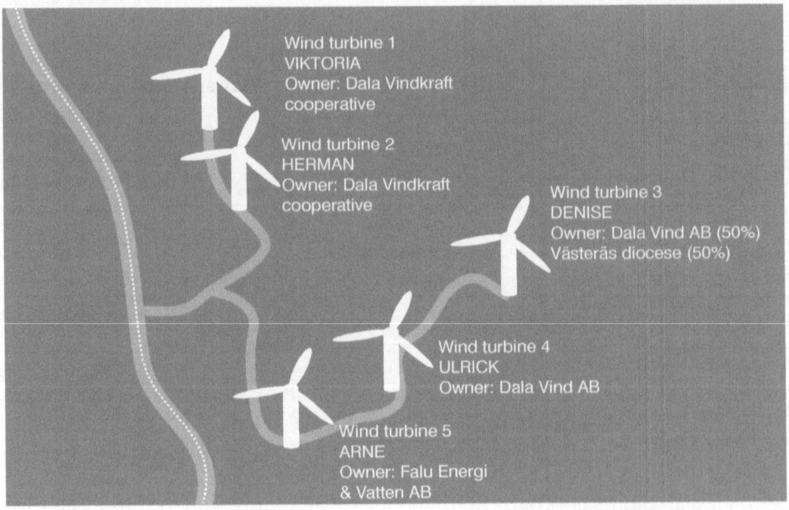

Figure 5.10 Högberget wind farm in Dalarna. This wind farm with five wind turbines of 2 MW has four different owners. Turbines 1 and 2 are owned by the windpower cooperative Dala Vindkraft ekonomisk förening; Turbine 3 is owned by Dala Vind AB and Västerås stift (Västerås diocese); Turbine 4 is owned by Dala Vind AB; and Turbine 5 is owned by Falu Energi & Vatten AB (a municipal company in Falun)

Each turbine has its own owner(s). However, a common "umbrella" company is formed wherein the owners of the single turbines have their shares. All incomes are put in a "basket", and all expenses are paid from the same basket. The cost for replacing a gearbox in one turbine is thus shared by all. This umbrella company that operates and maintains the turbines and distributes the revenues is often, but not always, managed by the project developer (a private company or a local utility).

This quite simple model can be used also for large windpower plants (see Figure 5.11). In this case, windpower limited companies, cooperatives, and other kinds of companies often own the individual turbines. However, in a windfarm with, say, five individually owned turbines, the revenues from the total production are divided into five equal parts. Windpower developers sometimes keep one or a few of the turbines in their own company as well (Wizelius, 2009/2012).

The windfarm Högberget in Dalarna in Figure 5.10 is one of many examples. Here are some others.

5.8.1 Ventosum AB

One example of a windpower plant managed by an umbrella company is Ventosum in Falkenberg. This windpower plant was developed by the municipal utility Falkenberg Energi in 1998 and consists of ten Vestas 660-kW turbines. These ten wind turbines have eight different owners; three are cooperative

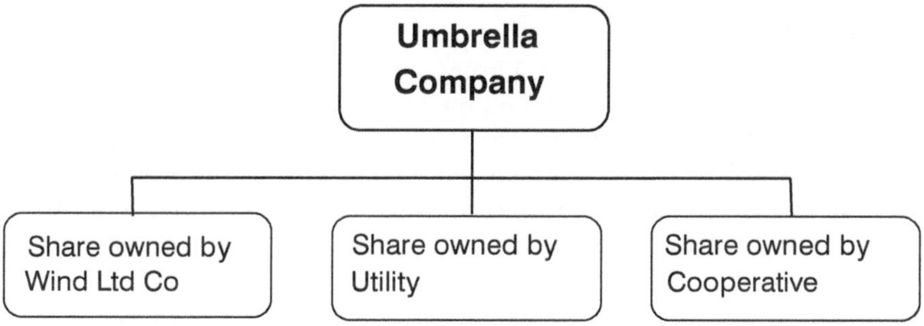

Figure 5.11 Umbrella company. This diagram shows the basic configuration of an umbrella company

associations, one local *Falkenbergs vindkraft ek för*, one national, *Sveriges Vindkraftkooperativ ek för*, and one regional farmers co-op, *Lantmännen ek för – Energi*. Four are windpower limited companies, and the housing enterprise in Falkenberg is the eighth of them (Wizelius, 2010/2012b).

These eight owners own shares in relation to their ownership in the umbrella company Ventosum AB. Ventosum does not own any shares in the turbines; it is a company formed to manage the windpower plant for the owners' best benefit. It supervises the windpower plant and takes care of the bookkeeping and service and maintenance of the turbines. It negotiates the power purchase agreement and manages the green certificates and everything else that needs to be done.

The wind turbine owners, who also own the umbrella company, just get their revenues. Incomes from all turbines are pooled together, and so are all costs. The profit that remains when costs have been deducted from the incomes is divided among the owners in relation to their share of the windpower plant. In this case, a company that owns one of the turbines will get 10 percent of the profit. To manage all this takes some expertise, not least to keep an eye on the certificate and power market to get the best deals. This expertise and negotiating power is hard for small wind turbine owners to get on their own.

5.8.2 Munseröd Vind AB

Munseröd Vind AB is an umbrella company that manages and operates the Munseröd windpower plant, with six 1.5-MW wind turbines from NEG Micon, which were installed in 2006. The shares are owned by the owners of these turbines, some private persons, a local Wind Ltd Company, Åsle Vind AB, farmers in the village Åsle, Slättens vind ek för (a cooperative association), and some private companies. It is operated the same way as Ventosum AB; however, in this company, the developer has no part, and the operation is managed by the members of the board (Wizelius, 2009/2012).

5.8.3 *Vindpark Vänern*

The offshore windpower plant Vindpark Vänern, with ten 3-MW wind turbines, is operated by two different umbrella companies, Vindpark Vänern Kraft AB and Vindpark Vänern Drift AB. They manage and operate five wind turbines each. The turbines owned by municipal companies have their own umbrella company, since rules for such companies differ from those that apply to private companies.

5.8.4 *Partnership contract model*

The project developer Siral forms a group of investors when the necessary permissions have been granted and before the procurement of wind turbines is started. This group of investors signs a cooperation contract. When the windpower plant starts to operate, they form a consortium that will manage and operate the windpower plant. Each investor is a member of the board, but the actual work is done by Siral.

This consortium does not own any wind turbines. It manages the operation, administration, taxes, and so on, and the surplus is paid to the owners/investors each month or whatever the board agrees on. The windpower plants are managed as an umbrella company; costs and revenues from all turbines are pooled and shared in proportion to the shares owned by each investor.

In recent years, some umbrella companies have been formed as limited companies, which is a more stable legal entity, although it takes more administration (see Figure 5.12).

Similar models are used by other project developers, such as Eolus Vind AB and Triventus AB, which have been in the windpower business since the 1990s. The financing is up to each investor to arrange, and this has to be in place when the partnership is established, before the windpower plant can be built.

5.8.5 *Leasing model*

When a large repowering project was made on Näsudden on Gotland, a new ownership model was developed: the *leasing model*. Wind turbines that were built in the 1990s were replaced by fewer but bigger wind turbines, reducing the number of turbines and increasing the production by a factor of four.

The repowering has been done in stages, and for each stage, a new umbrella company has been formed. The latest stage consists of ten 3-MW wind turbines, which were sold in parts of 1 percent each at a price of 3 million SEK/part. These owner shares can be used as collateral for loans in banks.

The umbrella company is formed as a limited company, which manages operation, service and maintenance, power trading, and administration. This company takes care of all costs and revenues. All contracts, for land lease,

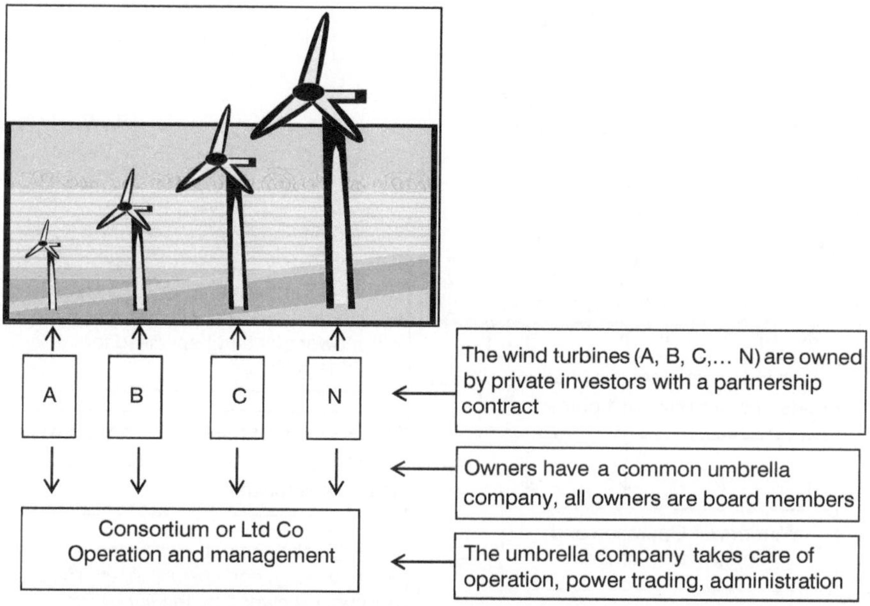

Figure 5.12 Partnership contract model. Investors make a partnership agreement and form an umbrella company that manages the operation and administration of the windpower plant

environment permissions, and so on, also belong to this umbrella company. The company does, however, not own the wind turbines.

The investors/owners have all signed a partnership contract, and they lease the wind turbines to the umbrella company. They own shares in the company proportional to their investments. Each month, the umbrella company pays rent to the owners with 98 percent of the net income; the remaining 2 percent is funded for repairs, and so on.

In 2013, three such companies are operating at Näsudden, with fourteen, thirty-eight, and eighteen part-owners, respectively. In this model, there are three interconnected levels: shares in the umbrella company, shares in the consortium (partnership contract), and share of ownership in the wind turbines (see Figure 5.13).

5.8.6 *Bixia ProWin AB*

Bixia AB is owned by ten different municipal companies or utilities and grid cooperatives. It was established in 1997 in response to the deregulation of the electric power market. In 2009, the company started an affiliated windpower limited company, Bixia ProWin AB. It offers some of its customers to buy shares in this company. Shares are offered only to company customers, not private consumers.

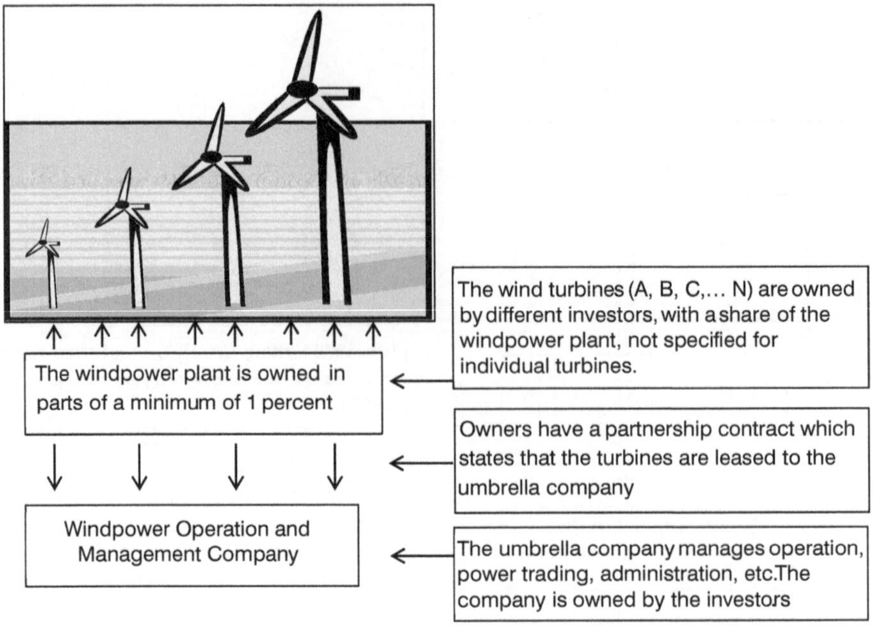

Figure 5.13 Leasing model. With the leasing model, the umbrella company leases the wind turbines from the owners and pays a monthly rent that corresponds to the net revenues

The ownership model is called Niondelsmodellen, *the ninth model*. Each wind turbine is divided into nine equal parts. One of these ninths is always owned by Bixia ProWin AB; the other eight parts are sold to companies that also are customers of Bixia AB.

One-ninth of the production of a modern wind turbine corresponds to about 700 MWh/year, which is an appropriate volume for many small and medium sized companies.

Several owners own one or more ninths of a wind turbine, which is part of a windpower plant with several turbines. Consequently, they own one-ninth of the costs for operation and maintenance and one-ninth of the power produced.

Each wind turbine is then operated as a partnership (in Swedish, *enkelt bolag*), which is a cooperation agreement (contract) between the part-owners.

Bixia ProWin AB takes care of the management of the windpower plants (i.e., operation, surveillance, power trading, administration, and so on). This is regulated by a management agreement (contract). Finally, there are two options for part-owners: one is to lease the part(s) of the wind turbine to Bixia ProWin AB, which will pay the annual net-revenues as rent, or to let Bixia AB transfer the power to the customer/part owners for net accounting (see Figure 5.14).

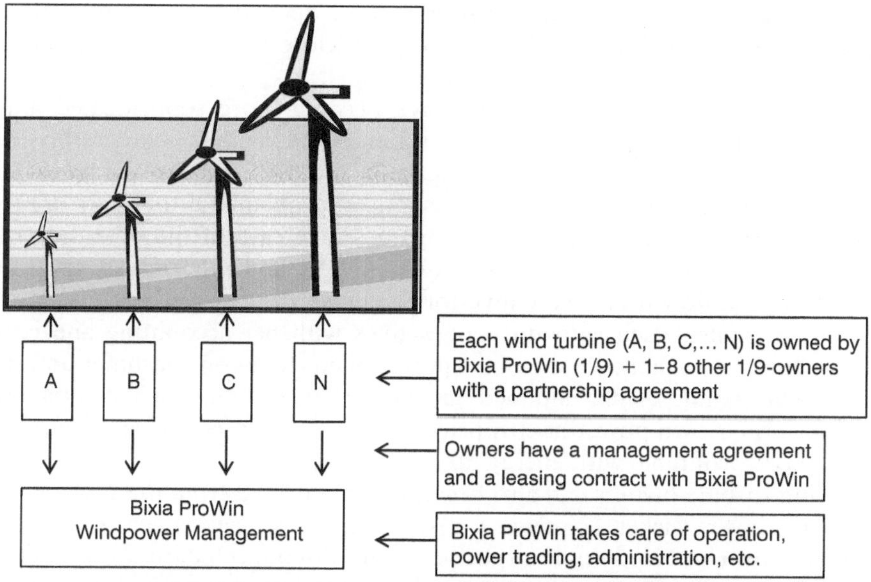

Figure 5.14 The Bixia ProWin AB ninth-model. This model is quite similar to the leasing model (Figure 5.13). Here, however, the power trader Bixia AB is part of the model; all power is traded by this company

The wind turbines in Bixia ProWin windpower plants are then managed by the umbrella company principle so that all production, revenues from certificates, and all expenses from the windpower plant are pooled together and shared by all owners in proportion to their ownership shares.

For the companies who are part-owners this model has several advantages. For those companies who have a power sale agreement, the costs for electric power and the revenues from the windpower plant follow each other, which will reduce volatility and risk. When the Nord Pool power price is high, revenues from the windpower plant will be high, and vice versa. Bixia ProWin AB calls this *energy-price neutrality*. Another term is *hedging*.

Companies with net accounting[3] (i.e., who use the power from their wind turbine shares themselves) do not have to pay any electricity tax.[4] All companies can make a depreciation of the investment up of to 33 percent annually.

5.9 Summary and conclusion

Many different ownership models for windpower are used in Sweden. As power companies did not show any interest in the development of renewables, this was done by grassroots, who initiated the first projects and had to find ways to finance them. Cooperative associations and limited companies, where capital for the investments were pooled by members and shareholders, were the only feasible way to do this, at a time when no banks would grant any loans for this new technology.

Deregulation of the electric power market opened up new innovative ownership models, which were developed in close cooperation with some local utilities, such as the net-accounting model.

During the years, these models have become more sophisticated and tailored for private investors, farmers, and small and medium-size enterprises, such as the partnership contract, the leasing, and the Bixia ProWin models.

The point of these models is to make it unproblematic for investors to be windpower owners. Umbrella companies take care of the operation of the wind turbines, manages costs, revenues, VAT and other taxes, and the monthly cash flow back to the investors/owners.

Other models such as wind cooperatives with net accounting and real estate companies that own their own power plants focus on consumer power. Private households, small and medium enterprises, and municipal companies produce their own power at cost price, and in some cases, the law makes it even more profitable with energy tax exemptions.

Wind turbine owners can also become disconnected from the Nord Pool electric power market, as power is produced at cost price. Even those who sell all power to Nord Pool at spot prices are disconnected, in the way that costs for power and revenues from windpower follow each other. This works as a hedge and neutralizes the volatility of the market.

Local ownership and community power are also promoted by many developers. From the start in the early 1990s and up to now, many windpower projects have been initiated by local actors. A group of people try to develop local windpower projects or get in touch with windpower development companies and ask them to do it for them. Many developers also offer people who live in the vicinity of a new windpower plant to become part-owners, either by direct investments or as members of a windpower cooperative.

The models with umbrella companies make it possible to finance windpower plants with a number of small investors. This model has also some obvious advantages for the project development companies, whether these are municipal utilities or private companies.

Ventosum is managed by staff from the municipal utility Falkenberg Energi. The wind farm Högberget is managed and operated by the regional Dala Kraft AB, owned by municipal utilities in the county Dalarna. The affiliated company Dala Vind AB, with the same owners, has developed the windpower plant. Other umbrella companies are managed by windpower project developers (i.e., the company that has developed and sold the turbines in the windpower plant).

Often the developer has no shares in the umbrella company; they just manage it for an annual fee from the owners. This is a very good business idea. By staying with the windpower plant they have developed during its entire lifetime, they will take responsibility for it, and this creates a relation of mutual trust between developer and buyers/owners.

To run a windpower project development company is a risky business in Sweden, at least if it is a small company with limited assets. The lead time for projects is hard to predict as well as the outcome. Owing to the complex procedure to apply for permission to build a windpower plant, a project can get stuck in the pipeline for years. The upfront costs are quite high, and to wait such a long time for some income from the work being done is hard to manage for small companies. By managing some umbrella companies, the developer gets a regular income and cash flow, which makes it possible to stay in business (Sirland, 2004).

Since 2006, some new large actors, power companies, and industrial companies have started to develop and invest in windpower plants. Some large industrial plants have installed them next to their plants and feed the power directly into their factories. Others build them on land they own, connect them to the grid, and sell the power on the market. This shows that windpower has become a competitive mainstream technology.

The development of ownership models for windpower has been very innovative and dynamic. The motive of different owners and for creating these different ownership models is the topic of Chapter 7.

Notes

1 The annual energy in the wind can vary by 20 percent compared to a long-term average. Members' power consumption tends to vary as well, owing to annual average temperatures.
2 This gives one grid fee instead of one for each apartment.
3 Bixia applies a one-hour balance for these customers.
4 Companies with net accounting do not have to pay energy tax; private members of wind cooperatives who use net accounting for their household consumption have to pay this tax.

6 Windpower ownership in Sweden

This chapter presents an overview of different actors who own and operate windpower plants in Sweden. There are several ways to sort ownership into categories. The choice depends on the focus and purpose of the classification.

In the annual reports that have published windpower statistics since the early 1990s, five different wind turbine owner categories are used, and these are the starting point for the analysis of the ownership distribution and development trends. These ownership categories are then subdivided and reorganized into some new categories, more relevant for the analysis of the topic of this thesis.

6.1 Windpower development in Sweden

When it comes to windpower development, Sweden is still behind its southern neighbors Denmark and Germany, countries who had a lead in windpower development from the 1980s (Denmark) and 1990s (Germany) (Wizelius, 2007).

However, the development pace in Sweden increased considerably in 2007 and 2008. Up to these years, the annual installations had been less than 100 MW. In 2007, there was a jump to 236 MW, 46 percent more than the preceding year and, in 2008, another 236 MW were added.

During the four-year period 2009–2012, the development pace increased even further, from 1,084 MW at the end of 2009 to a total installed capacity by the end of 2012 of 3,522 MW,[1] an increase of 325 percent (see Figure 6.1) (Vindstat, 2013).

When windpower development in Sweden started in 1990, the wind turbines were rather small. In the early 1990s, the turbines had a nominal power of 150 kW to 600 kW and were installed as single turbines or small groups of two to three turbines. In the last few years (since 2005), most wind turbines are large (1 to 3 MW) and expensive and are installed in large windpower plants (Wizelius, 2003/2007). The average size of wind turbines installed in 2012 was 2.3 MW (Energimyndigheten, 2013a).

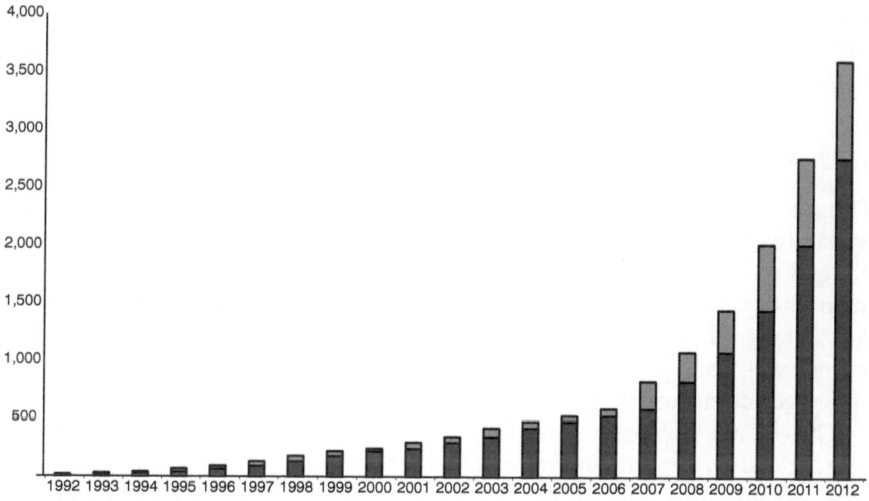

Figure 6.1 Windpower development in Sweden 1992–2012, MW. The light shading on top is annual added capacity; the light and dark show the total installed capacity. From 2007, annual installations have increased significantly (diagram based on figures from the Swedish Energy Agency)

6.2 Ownership categories

In the annual reports where windpower statistics have been compiled and published since the early 1990s, five different wind turbine owner categories are used:

1 private owners;
2 cooperatives;
3 windpower limited companies;
4 companies in other lines of business; and
5 power companies and municipal utilities.

These categories have been used since the windpower statistics started in 1991. There are also some additional categories: unknown (owner), no info, and only report (Elforsk, 1992–2012). The classification from official statistics above is used as a starting point for the description of windpower ownership.

6.2.1 Unclassified owners

The owners classified as "unknown" are among other foreign private or institutional investors. Those who are "not reported" (owners who do not report production figures) are not classified into ownership categories either.

The share of this category was around 3 percent in 2009 and 2010 but increased to 7 percent in 2011 and 13 percent in 2012 (Energimyndigheten, 2010–2013).

6.2.2 *Windpower statistics*

The statistics on windpower started in 1991, when an investment grant for windpower was introduced. To get the grant, the owner/operator was also obliged to report on monthly production and so on, and it was published in monthly as well as annual reports. The task to compile these reports was delegated to Vattenfall (Elforsk, 1992–2012).[2] When the promotion system was changed to the certificate system in 2003, this reporting was no longer compulsory.

These data are of general interest and also very valuable for project developers. However, some windpower companies no longer reported their productions and so on, as they considered such data to be business secrets. Another compulsory reporting system followed automatically with the certificate system. To get certificates, power producers have to apply and then report the production, and this is registered. With data from the certificate system, the Energy Agency compiled reports themselves, based on these (Energimyndigheten, 2013b). They also integrated data from the Vattenfall reports to get historical data. In the annual report for 2012, Vindstat AB has added all the wind turbines that do not report production data.

To get correct statistical data on ownership, the database from Vindstat has been used, and all entries on ownership have been checked. Then these "unknown" owners have been classified into relevant categories. Quite many were classified into wrong categories in the official reports and have been reclassified. To find unknown owners, the certificate database from the Energy Agency (Energimyndigheten, 2013b) has been used. With this database, some ownership categories have been sorted into subcategories for a more detailed analysis.

Many windpower projects in Sweden have a mixed ownership structure, so it is not always possible to find the exact shares of different owners. Umbrella companies, which do not formally own any wind turbines, are registered as owners. The actual owners may, however, belong to different ownership categories. These are classified as windpower limited companies, except in those cases where the owners of the specific wind turbines are known.

The same problem occurs with joint ventures, where, for example, one power company and one "other company" form a common windpower limited company. As the shares in the joint venture are not public and not always are fifty-fifty, these are classified as windpower limited companies. Figures may have some errors; however, not more than to give reasonably correct information on the share of different owner categories and development trends.

Some wind turbines that have been dismantled in re-powering projects are installed at new sites. Others are exported or dealt with as scrap. There is also a market for wind turbines on line, so that the original owner may sell some turbines to new owners. All of these transactions are not properly registered in the statistics. There may be some minor errors in the statistics owing to that. Windpower owners and how they are categorized are presented in Appendix D.

6.2.3 *Private owners*

There are some private persons who own wind turbines directly and not by part-ownership in companies. A large part of the members of windpower cooperatives, as well as shareowners in local windpower limited companies, are also private persons, but these are included in the categories windpower cooperative and windpower limited company, respectively.

Many of the private owners are farmers who have installed wind turbines on their own land or on leased land on sites where the wind resources are good. There are also private owners who are not farmers. The relation farmer–non-farmer is not specified in the statistics. There has been a steady increase of private owners during the period (see Figure 6.2).

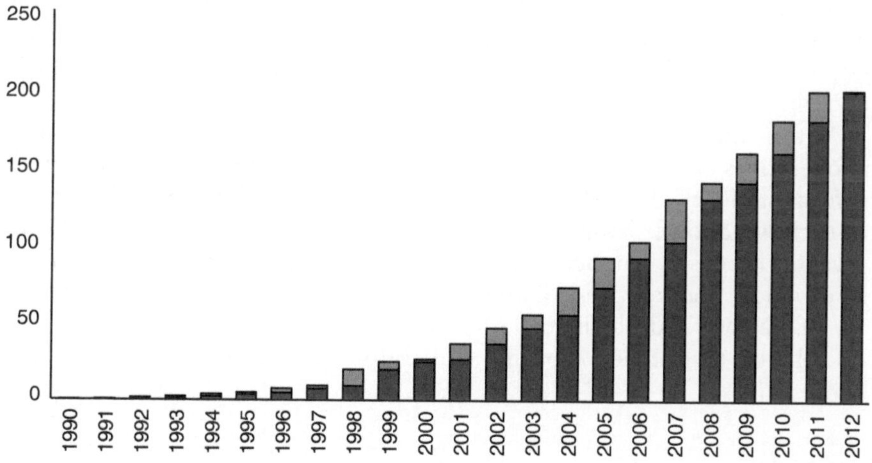

Figure 6.2 Private owners. Private ownership has increased in a regular way during the last fifteen years. In 2012, around 4 MW of new privately owned wind turbines were added, but this year also 3.5 MW were demounted in re-powering projects. The trend is stable, although the diagram indicates a break

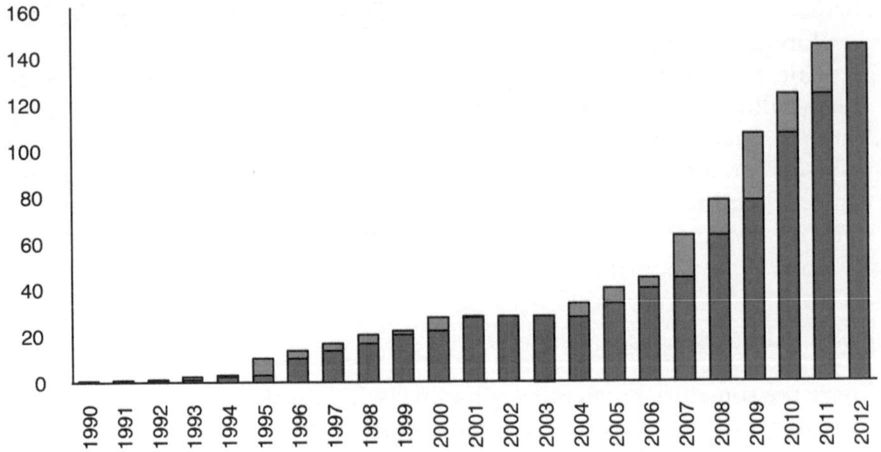

Figure 6.3 Windpower cooperatives. Windpower cooperatives show a growing trend from 2007 and tripled in six years. In 2012, only 2 MW of new cooperatively owned wind turbines were added (compared to 18.8 MW in 2011), but this year, also 2.5 MW were demounted in repowering projects. This may indicate a break in the growing trend for windpower cooperatives

6.2.4 *Windpower cooperatives*

The first windpower cooperative in Sweden was established in 1990. This category was a dominant owner during the early 1990s, just as in Denmark at the same time period. Members/share owners are not only private persons but also small and medium-size enterprises, and municipal and private real estate companies. From 2007, many new cooperatives have been established (Wizelius, 2010/2012a); (see Figure 6.3).

Within this category a distinction can be made between consumer cooperatives and producer cooperatives. This distinction is not made in the official statistics. To find this out, it is necessary to look into the statutes of each windpower cooperative, if it is not stated clearly elsewhere. Of the seventy-three wind cooperatives operating in 2012, at least three are known as producer cooperatives: Slättens vind ek för (30.5 MW); Bjärke vind ek för (0.8 MW); and Blekinge vindkraft ek för (2.4 MW). Around 23 percent (in relation to installed capacity) of windpower cooperatives are producer cooperatives.

6.2.5 *Windpower limited companies*

In the 1990s, several local limited companies were formed with shareholders from a small region who invested in wind turbines in the same region. These were based on the same idea as cooperatives but just organized differently. Some of these companies are now working on a national and

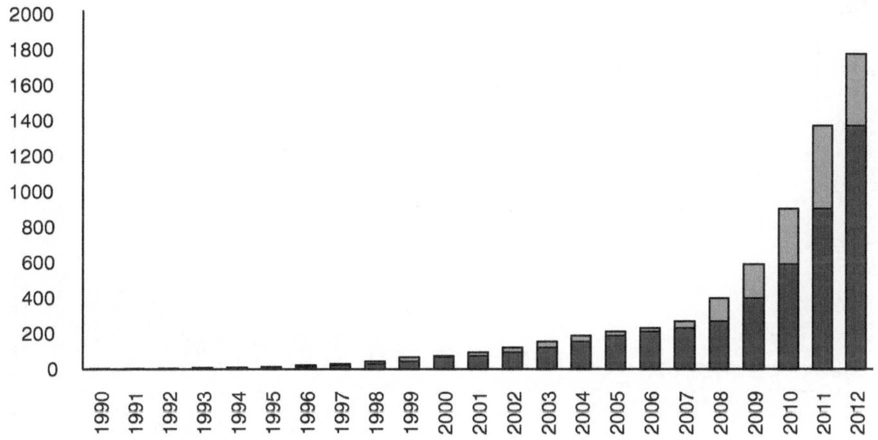

Figure 6.4 Windpower limited companies. These companies have increased their installed capacity significantly from 2008 and in following years

even international scale, developing and investing in wind turbines in other parts of Sweden and in other countries as well.

A new trend from 2005 was to establish windpower companies to be registered on the stock exchange. Arise Windpower AB is one of these new companies that develop, build, and operate windpower plants. Some municipal utilities also formed windpower limited companies, such as Dala Vind AB. More than half of the windpower plants in Sweden are owned and operated by windpower limited companies (see Figure 6.4).

In 2012, there were 294 windpower limited companies that owned windpower plants according to the official statistics. There are actually more; Åsle Vind AB mentioned in Section 7.3.2 is not included as it is part-owner in windpower plants operated by umbrella companies.

6.2.6 *Other companies*

This ownership category consists of companies with another purpose than power production. Many are farmers who operate their farms as limitedcompanies. Of the 150 other companies, sixty-one are agricultural companies and another nineteen are from agricultural business.[3]

During the last years, many real estate companies, private as well as municipal, have invested in their own wind turbines to use the electric power in the buildings they own (Energimyndigheten, 2013c).

Several large industrial corporations have also invested in wind turbines, some directly connected to their factories, others with normal grid connection. The paper pulp factory in Skutskär, owned by Stora Enso, installed five 2-MW turbines in their industrial harbor in 2009, with a direct connection to the factory (Sörgard, 2009).

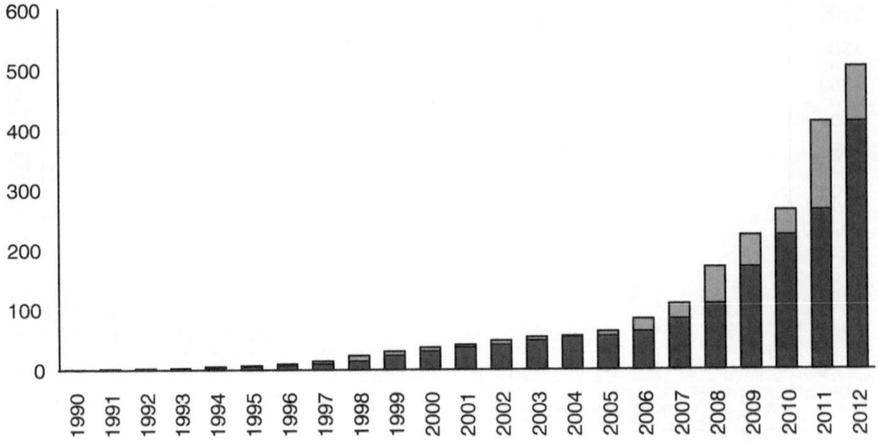

Figure 6.5 Other companies. Companies from business other than energy made large windpower investments in 2011–2012, and more are expected to come in the near future

In 2010, IKEA, the international furniture company, bought a windpower plant to supply electric power to its furniture stores in Sweden (IKEA, 2011).

This ownership category shows a sudden increase from 2011 (see Figure 6.5). This trend will continue, as many of these other companies have large windpower projects in the pipeline that will be connected to the grid in the next few years.

Many "other companies" are also members/owners in windpower cooperatives or part-owners in umbrella companies, which are sorted into the category of *cooperatives* and *wind limited companies*, respectively. The share of other companies is thus larger than the one given here. Data on this level are not easily available, so the analysis does not go further than registered owners in data from public statistics.

6.2.7 *Power companies and municipal utilities*

Power companies, such as Vattenfall and E.ON, own and operate power plants. It is the core business of these companies. Most of these power plants are conventional and also rather old hydro power plants and nuclear power plants and, during the last few decade, combined heat and power (CHP) plants have been added. Some of them also own and operate windpower plants, but in relation to the size of these power companies, the share of windpower is still marginal.

Several municipal utilities, such as Falkenberg Energi, Varberg Energi, and others had already developed windpower plants in the 1990s. From 2006, many other municipal utilities started to develop and invest in windpower plants.

Up to 2006, the capacity was rather modest. The jump from less than 100 MW to more than 200 MW in 2007 is due to the offshore windfarm at

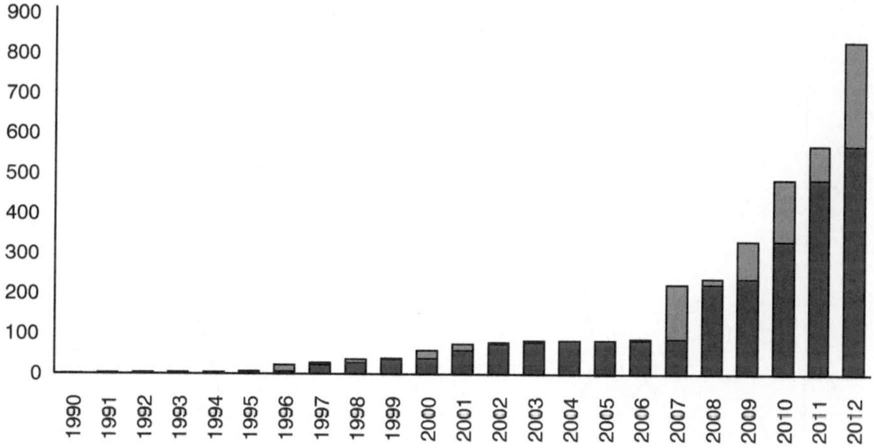

Figure 6.6 Power Company and Utilities. Power company and utilities started to make large investments in windpower in 2007 (Lillgrund offshore windpower plant) and more from 2009

Lillgrund owned by Vattenfall. From 2009, investments by power companies and utilities increased very rapidly (see Figure 6.6).

6.2.8 *Development trends*

All of these five ownership categories have a growing trend. This is quite natural when windpower development is increasing. Note, however, that the scales on the y-axis in these diagrams differ. More interesting is to see how the distribution between different ownership categories and their relative share of installed capacity has changed over time (see Figure 6.7).

The relative share of the total installed capacity of different owner categories has varied during this period. The share of wind limited companies has doubled. Power companies and utilities had a large share in the very beginning, but the share has decreased to around 20 percent. In 2011, their share started to increase. The share of private and cooperative ownership is declining. Other companies have had a share just below 20 percent during this period.

6.3 Other ownership categories

There are other ways to sort ownership into categories than the one used so far, taken from the official statistics. One obvious distinction is between utility generators, windpower plants owned and operated by power companies from the dominant energy system (DES), and those from other independent windpower producers.

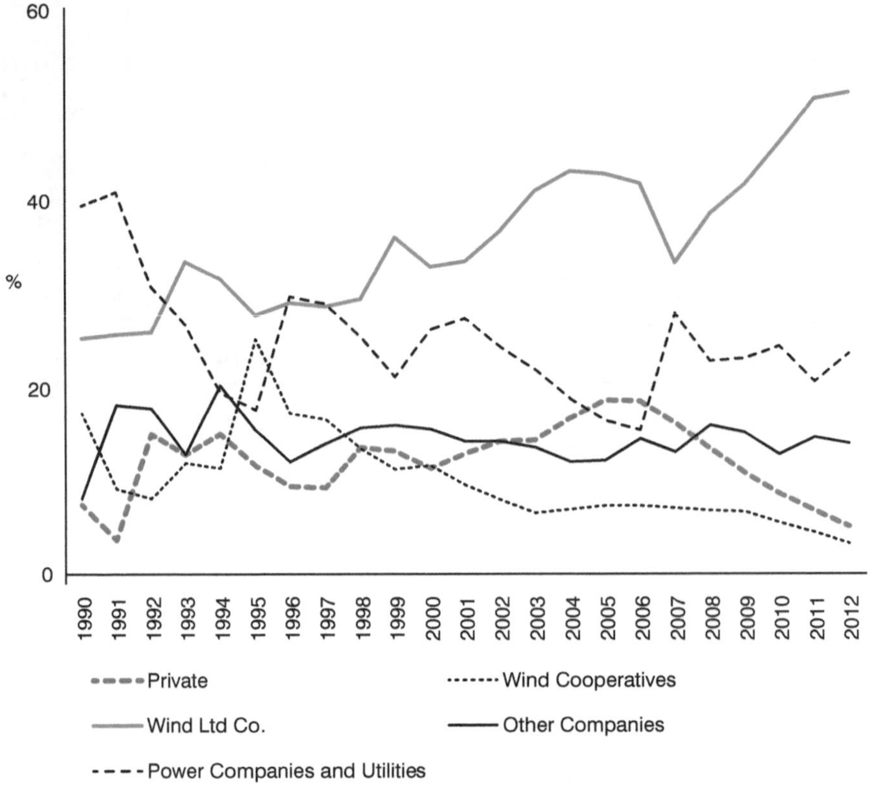

Figure 6.7 Share of different ownership categories, 1990–2012

Another distinction could be made between those windpower plants that could be considered as "community power" projects, with local investments, ownership, and benefits, and other windpower plants. Some windpower plants can be defined as consumer-owned power production. Wind cooperatives belong to this category, but also do some of the windpower plants owned and operated by other companies.

Within the category of other companies, those companies that are farms or have a direct connection to the farming business can be sorted out to find out in what respect windpower can be considered to be a farming business. Some of these categories are overlapping; for example; some but not all windpower cooperatives are also community power projects.

6.3.1 Non-utility generators

Independent power producers (IPPs), also called non-utility generators (NUGs), are defined as producers of electric energy that are not public utilities but make electric energy available for sale to utilities or the general public. As all power producers on a deregulated and unbundled market by

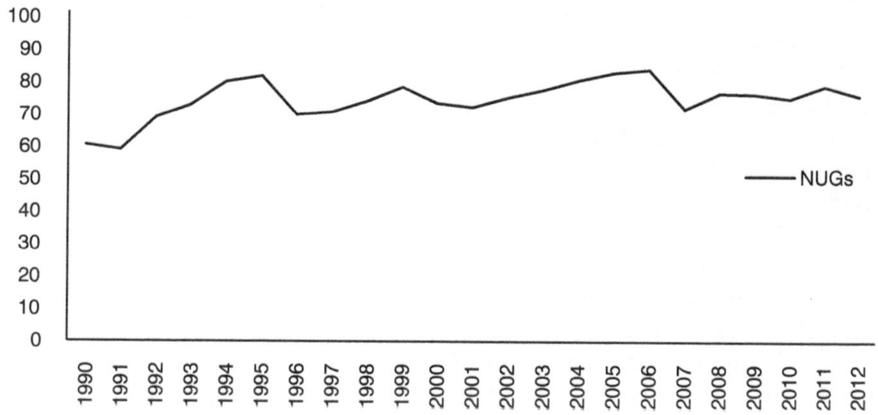

Figure 6.8 Non-utility generators' share, 1990–2012, percent

definition are IPPs, I use the concept *non-utility generator* in the following analysis and discussion. A distinction is thus made between power companies that belong to the DES and other companies that invest in windpower.

With the classification of ownership used in Sweden, the categories (1) private, (2) cooperative, (3) windpower limited companies, and (4) other companies in Section 6.2, are NUGs. During the period 1990–2012, most wind turbines have been owned and operated by NUGs (see Figure 6.8).

The share of NUGs has varied from 60 percent (1990–1991) to a maximum of 84 percent in 2006 and, in 2012, the share of NUGs was 76 percent.

6.3.2 *Transnational versus local/regional power companies*

In the category of utility generators (i.e., ownership category e: power companies and utilities) there is a distinction to be made between power companies and municipal utilities. The latter are locally owned and can be regarded to represent *community power,* while power companies are actors on an international market.

Transnational power companies

The state-owned power company Vattenfall has from the end of the 1990s been an actor on the international market (Högselius & Kaijser, 2007).

In 2012, Vattenfall had 900 wind turbines in operation. Of these, 120 were operating in Sweden, and the remaining 780 turbines were installed in Denmark, Germany, the Netherlands, and the United Kingdom. The largest windpower plants are offshore: Horns rev in Denmark, Lillgrund in Öresund, Egmond an Zee outside the Dutch coast, and Thanet in southeastern England (Vattenfall, 2013). Most investments have been focused on countries with feed-in tariffs or with long-term power purchase agreements for offshore windpower.

Power companies from other countries have, on the other hand, entered the Swedish power market. Most of them are companies from neighboring countries in the same Nordic power system, which already were operating in Sweden since the 1990s. These started to show some interest in windpower plants from 2007 onward (i.e., the year that windpower development finally took off also in Sweden).

E.ON, based in Germany, entered the Swedish market in 2001 after the purchase of Sydkraft AB, a power company owned by a number of municipalities in southern Sweden.

E.ON Vind operates a number of windpower plants, most of them with three to ten wind turbines. The oldest one is from 1996 with twelve turbines on an artificial island outside Landskrona; one from 2001 in Malmö harbor; and all the rest from 2011–2012. A medium-size offshore windpower plant outside Öland, with sixteen wind turbines and 48 MW capacity, will be taken into operation in 2013 (E.ON, 2013).

The Finnish state-owned power company Fortum invested in the municipal utility in Stockholm in 1998 and took over the company in 2001. Fortum did not have any significant activity in windpower development before 2010, when it formed the joint venture Blaiken Vind AB with Skellefteå Kraft, a project with 100 wind turbines at Blaiken in Västerbotten county[4] (Fortum, 2013).

The Norwegian state-owned power company Statkraft has a much smaller share of the power production in Sweden than Vattenfall, E.ON, and Fortum: only about 1.5 percent (2 TWh/year). Statkraft's first windpower plant, on the island of Smøla in Norway, was in operation in 2002.[5] Statkraft has set the target to develop 1,100 MW windpower on land and 150 MW offshore before 2015.

In Sweden, Statkraft got involved in windpower in 2006 when it formed the joint venture Statkraft SCA Vind AB with the large Swedish forestry and paper pulp company SCA. In the southern part of Sweden, Statkraft cooperates with Södra Skogsägarna, a farmer-owned forestry and paper pulp company. Statkraft is developing forty windpower projects in Sweden in cooperation with these two partners. Statkraft will within a few years be one of the largest windpower producers in Sweden (Statkraft, 2013).

DONG Energy was established in 2006 by the merger of six Danish energy companies, and 80 percent of the company is owned by the Danish state.

Dong Energy had 600 MW windpower on line in 2012, and more projects are developed in the United Kingdom, Germany, Poland, Denmark, and Norway. In Sweden, Dong owns one windpower plant in Storrun in northern Sweden with a capacity of 30 MW.

All these five companies are parts of the DES and are operating in all sectors of the power system: power production, grid operation, and power trade. Their interest in making investments in windpower plants from 2007 onward shows that windpower now has gotten the market acceptance in the power business.

Local regional power companies and municipal utilities

Several municipal utilities have been engaged in windpower development since the early 1990s, such as Falkenberg Energi, Varberg Energi, Borgholm Energi, and Kalmar Energi. They have also promoted windpower cooperatives for their customers. From around 2005–2006, several other municipal utilities started to invest in windpower: Umeå Energi, Skellefteå Kraft, and Karlstad Energi, to name a few. Others, such as Robertsfors, invested in windpower cooperatives. Telge Energi chose another role; it buys and resells power from wind and other renewables but has also started to offer its customers to invest in windpower cooperatives.[6] In the county Dalarna, several municipal utilities formed the company Dala Vind AB in 2006 to develop, own, and operate windpower plants; and some years later, a similar company, Samkraft AB, was formed by twelve municipal utilities in the counties of Gästrikland, Hälsingland, Uppland, Västmanland, and Södermanland.

Bixia AB is formed by several municipal utilities in southern and middle Sweden and offers local companies to invest in windpower plants. Several municipal real estate companies have also invested in windpower plants, but these are sorted into the category "other companies."

Many municipal utilities are thus actively engaged in the transition to a renewable energy system. Their activities are in most cases restricted to the municipality or a larger region covering several counties. They produce power for local consumption.

6.3.3 *Power companies versus municipal utilities*

When the category of power companies and utilities is subdivided into transnational companies[7] and local/regional utilities,[8] it turns out that the transnational companies' share and installed capacity has varied quite a lot during the period 1995–2012 (see Figure 6.9).

In 2012, transnational companies owned 46 percent of the installed capacity of power companies and utilities and twenty-nine utilities and local power companies (54 percent). Transnational power companies had about 11 percent of the total installed capacity in Sweden in 2012.

6.4 Community power

Another distinction can be made between windpower plants that could be considered as *community power* projects and other windpower plants. Community power was discussed in Section 4.2.4.

According to the World Wind Energy Association, a project can be defined as community power if at least two of these three criteria are fulfilled:

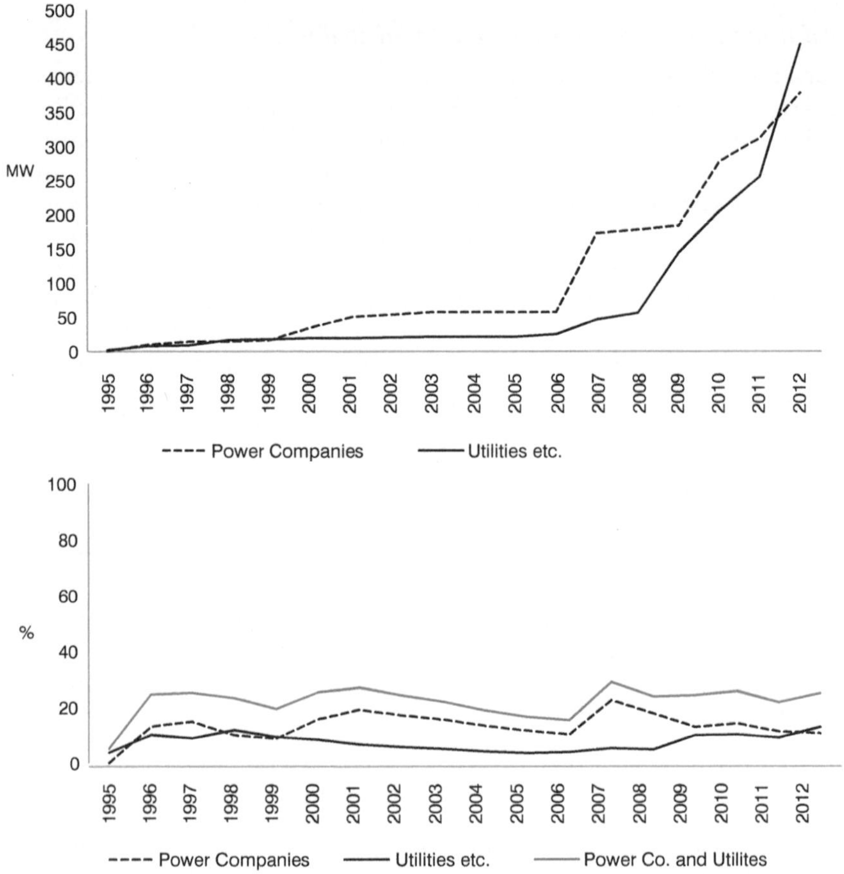

Figure 6.9 Power companies vs. municipal utilities, 1995–2012, MW and percent. Before 1995 there were no commercial wind turbines operated by power companies

1 local stakeholders own the majority or the entire project;
2 voting control rests with the community-based organization; and
3 the majority of social and economic benefits are distributed locally.

With two of these three criteria fulfilled, the major part or all of the social and economic benefits are returned to the local community, according to WWEA (Martin, 2011).

Local stakeholders can be farmers, cooperatives, independent power producers, financial institutions, municipalities, schools, and so on.

With the definitions of ownership categories, projects owned by municipal utilities and local/regional power companies are community power projects.

So are those private-owned wind turbines that are sited close to their owners; wind limited companies with local windpower plants and local shareholders; and windpower cooperatives with a majority of local members.

None of the five categories in the official statistics can be defined as community power as a whole, but within all of them, there are projects that can fulfill the criteria for community power projects.

The wind turbines and owners in the database have been classified to figure out the share of community power projects in each of these categories and of the total installed capacity.

For local ownership and community power, the geographical site of the owners is the basis for the classification. However, the definition of local is problematic. For a farmer, *local* would mean in the backyard; for a municipal utility, a windpower plant within the municipality, and so on. *Local* has different meanings for different owner categories. Therefore, different definitions of *local* are used for different ownership categories (see Table 6.1).

Of the windpower limited companies, 473 MW are local companies, operating one or a small group of wind turbines; 55 MW are regional, operating several windpower plants in one or several counties. Some 100 MW are operated by umbrella companies, where it is hard to get the details on part-owners. However, in most, there is a large share of local investors, as this is the business idea of the project developer. With a conservative estimate of 50 percent, this will add 50 MW of local ownership.

In the category *other companies*, farming companies can be sorted out. Farming companies own 56.6 MW. There are also companies in the agricultural business, which add another 38.7 MW. Among the category *private*, another 40 MW can directly be identified as farmers, but the share among private owners is likely much higher. This adds up to 135 MW, but many farmers are also part-owners of windpower limited companies

Table 6.1 Share of "local" ownership

Owner category	*Definition of local*	*MW*	*Local in % (MW)*
Power companies and utilities	Region/one or more counties*	450	76
Private**	Municipality	136	66
Wind cooperatives	Municipality/county***	115	80
Wind limited companies	Municipality/county	578	32
Other companies	County	121	24
Total		1400	40

* The marketing area defined by the local utility/power company.

** Rough estimate. First. private owners who are not local have been subtracted. Of the remaining, 75 percent are assumed to be local (it has not been possible to check all owners).

*** Some local windpower cooperatives have a few members from other regions.

and windpower cooperatives. Windpower is thus an important business for farmers, who are important actors in the windpower business.

In 2006, the government introduced the concept of "Vindbruk," which means wind farming, to stress this point in the government bill on windpower (Swedish Government, 2006). The national farmers' association LRF has also realized this and has published information material for their members about windpower investments, land lease contracts, and a windpower manual (Wizelius, 2009/2012).

When all windpower plants in Sweden are classified according to these criteria, it turns out that at least 40 percent are community power projects.

6.5 Consumer windpower

Some windpower plants can be defined as consumer-owned power production. Wind cooperatives belong to this category, as do some of the windpower plants owned and operated by other companies. Among these, there are private and municipal real estate companies, housing companies, and large industrial corporations.

Consumer power does not necessarily mean that the power produced is fed directly into the owner/consumers' buildings or factory; this can be handled by power traders and the distribution system. Also, if the power produced is sold to the Nord Pool, this can in some cases be considered as consumer power, as this works as a hedging of power prices for the owners. This will be described in Chapter 7, Windpower Owners' Motives.

The share of consumer windpower started to grow from 2005, when it was 3.4 percent, to 7.5 percent in 2009, when it leveled out. In 2012, the share of consumer windpower was 5.9 percent. However, considering the plans of Statkraft SCA Wind AB, which is building 915 MW windpower, this share will increase in coming years (see Section 7.4.1 and Table 7.1). SCA has a 40 percent share in this joint venture (i.e., 366 MW) and, added to the 203 MW consumer windpower in 2012, the share of consumer windpower will double.

Large investments by industrial companies add to the wind cooperatives that have been a significant ownership category since the early 1990s. Another group of consumer windpower owners are private and public real estate companies, from the category of other companies (see Figure 6.10 and Appendix E, list 3).

6.6 Summary

In 2007, there was a significant increase of the annual installations of windpower in Sweden, a take-off. Large investments were made by all five owner categories; all show a growing trend in installed capacity (MW). However, some categories increased relatively more than others (see Figure 6.7).

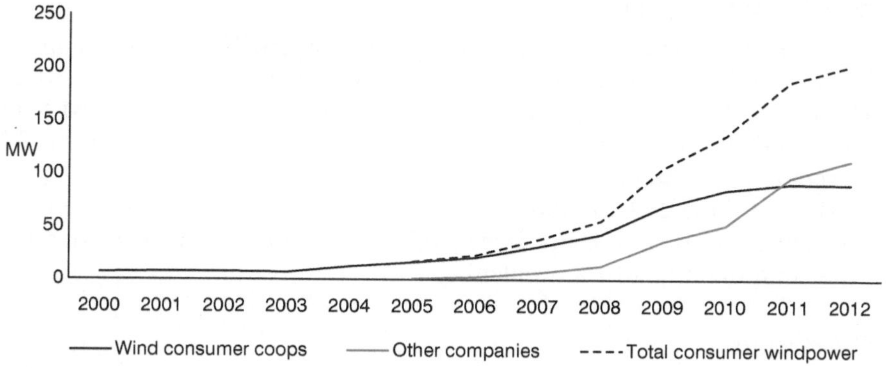

Figure 6.10 Consumer windpower, MW. The development of windpower cooperatives has stagnated, but the investment of other companies is increasing

6.6.1 *Shares of original owner categories*

During the period 2007 to 2012, after the take-off of windpower development in Sweden, the distribution among the five owner categories used in the official statistics changed.

Private owners

The share of private owners decreased. In 2007, the share was 17 percent of the installed windpower capacity; in 2012, this was reduced to 6 percent. However, there are many private owners "hidden" in umbrella companies, which are sorted into the category of wind limitedd companies. This is hard to quantify. Still, this owner category has most likely decreased in relative terms.

Windpower cooperatives

The share of windpower cooperatives decreased, although the installed capacity owned and operated by wind cooperatives grew from 44 MW in 2006 to 144 MW in 2012. In 2006 and 2007, the share was 8 percent of the installed windpower capacity; in 2012, this was reduced to 4 percent. In 2009, many cooperatives had problems to recruit members/investors, owing to a new interpretation of tax rules for wind cooperatives (see Section 7.6.4). Several new cooperatives were never realized. However, the number of members in windpower cooperatives increased from 19,300 to 25,500 from 2009 to 2012 (Wizelius, 2010/2012a).

Windpower cooperatives have private households as members, but also many small and medium-size enterprises as well as private and municipal housing companies (i.e., members from the category *other companies*).

Wind limited companies

Wind limited company is the dominant owner category. This kind of company owns more than half of the installed windpower capacity in Sweden in 2012. The installed capacity increased from 267 MW in 2007 to 1,775 MW in 2012, and the share of ownership in the same period increased from 34 percent to 51 percent.

Most of these 294 windpower limited companies are small private local companies, but there are also large public ones who own many windpower plants in different parts of Sweden.

Other companies

The share of other companies increased. The installed capacity owned by other companies grew from 109 MW in 2007 to 503 MW in 2012. In 2007, the share was 14 percent of the installed windpower capacity; in 2012, this had grown slightly to 16 percent.

Of the 150 other companies in this category, 80 are from the agricultural sector. There are also several real estate companies, municipal as well as private, in this category, and some large industrial companies.

Power companies and utilities

The share of power companies and utilities decreased. The installed capacity owned and operated by power companies and utilities grew from 224 MW in 2007 to 830 MW in 2012. In 2007, the share was 28 percent of the installed windpower capacity; in 2012, the share was reduced to 24 percent.

6.6.2 *Shares of new owner categories*

Windpower owners were sorted into some new categories to explore other trends in the ownership distribution: utility generators, non-utility generators, transnational power companies, and local/regional power companies. Finally, owners who could be classified as “community power” and “consumer power” were sorted out.

Non-utility generators

The first four of the five ownership categories above were sorted into the new category of non-utility generators. It turned out that most windpower in Sweden has been and still is developed by such non-utility generators; their share has varied from 60 percent to 84 percent and was 76 percent in 2012 (see Figure 6.8). A breakdown of the category of power companies and utilities into transnational power companies and local regional ones showed that the share of transnational power companies, during the growth period 2007–2012, has decreased from 22 percent to 11 percent, although the installed capacity more than doubled, from 173 to 378 MW.

Community power

Windpower plants were then sorted with the criteria of local ownership, giving local benefits that can be considered as "community power." This was done within all the five ownership categories used in official statistics. There are no variables to distinguish local ownership in the database, and there are many projects and also umbrella companies, with a mix of local owners and owners from other places. Also, different definitions of "local" are used for different owner categories. This resulted in an estimate that at least 40 percent, and probably up to 50 percent, of the windpower capacity in Sweden can be defined as "community power."

Consumer power

Consumer power can be defined as investors (private households, SMEs, housing companies, industrial companies) who invest in windpower plants to get electric power for their own consumption, directly or indirectly (price hedging). The share of consumer windpower grew from 3 percent in 2005 to 7.5 percent in 2009 but leveled out and decreased slightly to 6 percent in 2012. The share of consumer power is expected to increase quite a lot in coming years.

6.6.3 Development trends

The trends for these old and new ownership categories show some interesting patterns. The share of small and local investors (private, wind cooperatives) seems to be declining. On the other hand, the number and the investments of windpower limited companies is growing. These small and local investors may have made their investments in such companies instead (i.e., moved to another ownership category). The large share of community power is an indication of this.

The growth of the number and capacity of windpower limited companies is impressive. There are a few new large national companies among these, such as Arise Windpower AB (268 MW); Bliekevare Vind AB (55 MW); and Rabbalshede Kraft AB (63 MW), and some that have been operating since the 1990s such as Slitevind AB (21 MW) and Eolus Vind AB (50 MW), but most of the capacity is still owned by small local companies.

Somewhat surprising is that the share of power companies and utilities is declining. For the transnational companies belonging to the DES, the decrease is even bigger, and they own no more than 11 percent of the installed windpower capacity in Sweden in 2012.

Finally, the share of community power is estimated to be at least 40 percent and likely even 50 percent. The share of consumer power has leveled out but can be expected to increase in coming years.

Notes

1 Figure from Vindstat AB and the database used for analysis in this chapter. The Energy Agency figure is 3,607 MW and Svensk Vindenergi 3,750 MW. These organizations use different criteria for when a wind turbine is included in the statistics.
2 From 2011 by an independent private company, Vindstat AB.
3 Agricultural equipment, sawmills, machinery, retailers of vegetables, and so on.
4 In operation in 2013, so this windpower plant is not included in these statistics.
5 This is a very large windpower plant with sixty-eight wind turbines, a capacity of 150 MW, and an annual power production of 450 GWh.
6 Telge Energi Vindförening owns two wind turbines. The first one started in 2011, the second in 2012, but they are part of umbrella companies and not visible in the statistics.
7 Vattenfall, Dong, Statkraft, E.ON, Fortum.
8 Public and private local/regional power company/utility (see Appendix D).

7 Windpower owners' motives

In Chapter 5 and Chapter 6, it was shown that there are many different owner categories and ownership models. These different categories can have quite different motives for investing in windpower plants. Power companies, industrial corporations, municipal utilities, and farmers may also have different reasons for owning wind turbines.

In this chapter, the focus is on research question 4: *Why are these ownership models used?*

After the ownership models described in Chapter 5 and the quantitative description of the distribution of different ownership categories presented in Chapter 6, we move on to a qualitative analysis, based on interviews with twenty windpower owner representatives (see Appendix A), and on public documents and other information from company Websites. The motives of these different actors and investors are analyzed one by one.

7.1 Private owners

The first owner category in the official statistics is private persons/farmers. Windpower owned by private persons has increased from 57 MW in 2003 to 205 MW in 2012. During the same ten-year period, the relative share of private owners has decreased from 15 percent to 6 percent, a clear downward trend.

7.1.1 Gotland farmer pioneer

The farmer Åke Ahlsten from Gotland made his first investment in windpower in 1993, when he became a member of the wind cooperative Österudd, with three 150-kW turbines at Näsudden in southwestern Gotland.[1] The following year he invested in another wind cooperative, Klasården, in the same area. In 1995, he joined forces with seven other persons and bought a 500-kW turbine, also on Näsudden, which was named Grupp 8 (Group 8). This turbine is a business venture, and the investment is, from a tax and business point of view, treated like an investment in a farming machine and a part of the part-owners businesses (six are farmers, two have other types of companies).

For Åke Ahlsten, the investments in windpower have been just like any other investment in the farming business he operates. The economic calculations have proved to be almost correct. The calculated productions were a bit optimistic, but on the other hand, the return on investment has been good enough, between 10 and 12 percent, which is significantly more than bank interests. For me, it has been profitable to invest in windpower, Åke Ahlsten concludes (Wizelius, 2009/2012).

7.1.2 A wind farmer from Falköping

Per Ericsson owns and operates a farm in Falköping municipality, situated in one of the best farming areas in the middle of southern Sweden. It is far from the coast, but the landscape is open, with good wind resources. There are also quite a few wind turbines operating in this area: 40 turbines, 43 MW (Energimyndigheten, 2013b).

Per Ericsson installed an 850-kW wind turbine on his farm in 2002. He developed the project, applied for building permission, grid connection and so on himself but had a partner with some experience of windpower to share the investment and formed a company, which owns the wind turbine.

He had considered installing a wind turbine long before that, but then the ends did not meet, he explains. In 2002, the economic conditions were finally good enough. He had two options at that time—to invest in a new pigsty or in a wind turbine—and chose the latter. This was a good choice; the turbine has made good profits, he concludes.

He also manages the sale of power and certificates himself, to get the best price. Before he installed the wind turbine, he made daily checks on the Internet of the world market prices for fodder to his pigs and other commodities.

Concern for the environment has also been in the balance. To stop climate change is also in the interest of farmers, he explains, and adds that windpower development in this region has increased the market price of farming estates due to better roads and a stronger power grid, which have been built for the transport and grid connection of wind turbines.

Since the wind turbine he installed at his farm has lived up to expectations, Per decided to invest more. He bought a second-hand but operating 750-kW wind turbine in the wind farm Mungseröd, on the west coast of Sweden, in 2005. He has also been a member of the wind cooperative Slättens Vind ek. för. several years before he invested in his own wind turbines (Wizelius, 2009/2012).

7.1.3 Motives of private investors

Farming is a business, and it is not surprising that investments in windpower are evaluated just like any other investment, which both Åke Ahlsten and Per Ericsson obviously do. They have considered this a good business opportunity, and when it turned out to live up to expectations, they both have made more investments. It is also interesting to note that both of them first were members of wind cooperatives and likely learned to trust this then-new technology and to do the economic calculations necessary to evaluate commercial investments in wind turbines.

7.2 Windpower cooperatives

The first windpower cooperatives in Sweden were established in the early 1990s. Members of windpower cooperatives own their own power plant. In a consumer cooperative, each member's share should correspond to their own power consumption (this rule defines the cooperative as a consumer cooperative, so that the revenues from sale of power are not taxed as a company profit).[2] In other countries, all windpower cooperatives are producer cooperatives (i.e., cooperative business ventures). The difference between producer cooperatives and limited companies is that in the cooperative each member has one vote, while in a limited company, influence is in relation to the number of shares (i.e., the size of each part-owner's investment).

Up to the deregulation of the electric power market in 1996, all cooperatives had to sell their power to the local utility, and revenues were then shared and distributed once a year. The advantage was that there was no tax on profits, which a limited company would have had to pay.

After the deregulation of the electric power market, some cooperatives developed new models, so that members could get the power at self-cost price instead of paying their ordinary bills and getting annual refunds from the coop. The next step in the development of wind cooperatives was that it was used as a business model by large windpower developers working on a national scale. Some municipal utilities also supported wind cooperatives and helped them to utilize these models for net accounting.

Windpower cooperatives can be defined as consumer-owned power production. There has been some discussion on interpretations of the electric power law and tax rules. Questions have been raised by the dominant energy system (DES) power companies, such as Vattenfall, and the tax authorities have made interpretations of how the laws should be applied. This will be discussed in Chapter 8.

Windpower cooperatives owned around 28 MW in 2003, and this figure increased to 144 MW in 2012, when there were seventy-three windpower cooperatives in Sweden. From 2009 to 2012, the number of members increased from 19,300 to 25,500 (Wizelius, 2010/2012a). Still, the share of windpower cooperatives decreased from around 7 percent down to 4 percent.

7.2.1 Driving spirit in Lysekil

In Lysekil on the Swedish west coast, there are three wind cooperatives in operation, and Christer Larsson is the chairman of the board in all of them;

on top of that, also the chairman of the board in a fourth wind turbine owned by a local limited company, *Vindinvest i Lysekil AB*. The first two cooperatives, *Lys-vind* and *Si-vind*, were established 1995, *Sivik III* and *Vindinvest i Lysekil*, were established in 2000.

Christer Larsson's interest in windpower started in the 1980s, when he participated in a study circle with a group that built some small wind turbines. In 1994, he invited the project developer *Vindkompaniet* to Lysekil to give a lecture on commercial grid-connected wind turbines. He then applied for permission to build two wind turbines at the municipal waste disposal site, and the first two cooperative wind turbines have been operating there since 1995. Five years later, two more wind turbines were installed in the same area.

It was easy to recruit members to the first two windpower cooperatives, according to Christer Larsson. They had some information meetings, and then the shares were sold in just two to three days, he remembers. When the third cooperative was started, most local people in Lysekil already had shares in the first cooperatives, so the new association marketed their shares in a wider region and in some parts of Gothenburg as well.

In the first wind coops, all local windpower enthusiasts became members. In the second and third cooperatives, some other categories of people joined, some who were curious and wanted to have a few shares but also some who joined only for economic reasons: to make good business investments.

Christer Larsson finds a wind cooperative to be the most interesting ownership model since it makes consumers produce their own power. It is also a good business for the state, since members do their investments with taxed money, but to be member of a cooperative is also a *political statement*; members prove that they want to create a sustainable power system, according to Christer Larsson (Wizelius, 2010/2012a).

7.2.2 Kvarkenvinden: A fast-growing wind cooperative

In the summer of 1998, a Bonus 600-kW wind turbine started to operate in Holmsund harbor at the Baltic Sea coast in northern Sweden. It was built by the cooperative association Kvarkenvinden. In 2010, this cooperative had seven wind turbines on line, all with 2- to 3-MW nominal power, producing 38 GWh annually.

The fast growth of this cooperative can be explained by two factors: entrepreneurial skill and good cooperation with the local utility. However, to the end of the 1990s, the utility was opposed to windpower. This changed after an election, when some new politicians were appointed to the utility board.

Kvarkenvinden and Umeå Energi created the net-accounting model (see Section 5.4.2). For the power from the wind turbines, members pay only the cost of operation and maintenance. There is, of course, also a capital cost for the investment to buy the shares. These shares are not repaid, but a member who leaves the association will get this money back at nominal value (i.e., without any interest). The price for the windpower is around 10 to 12 öre/kWh + VAT. This is very low compared to the average market-price, which is in the region of fifty öre/kWh (price variations are high, though). The price for the wind-generated electric power is low and also predicable.

To be a member of Kvarkenvinden is good business and an attractive option. Kvarkenvinden has always had a long line of prospective members and the capital necessary to invest in new wind turbines.

The members are not only private households. Most of them, according to the number of shares, are small and medium enterprises, and also some municipal companies (Wizelius, 2010/2012a).

This cooperative association was initiated by a local entrepreneur, who owns and operates several tenements in the Umeå region, and was organized by a group of university teachers and students and environmentalists with ambitions to create a sustainable energy system. So the motives to start this cooperative were idealistic, but today the motive to become a member is economic: members save significant amounts of money on their electric power bills (Wizelius, 2010/2012a).

7.2.3 Varbergsvind: Initiated by a municipal utility

Ringhals is the site for one of the three operating nuclear power plants in Sweden and of the large paper pulp factory Värö bruk. Between these two giant industrial plants in Varberg municipality, some eighty km south of Gothenburg, there is a group of three wind turbines. Half of them are owned by the municipal utility Varberg Energi, the other half by the windpower cooperative Varbergsvind (they own one turbine each and share the third one fifty-fifty).

The utility was an early starter in the windpower business and built its first wind farm in 1991, with seven Vestas 225-kW turbines, and a second wind farm with three 600-kW turbines in Tvååker in 1995. In 1999, the utility got an offer to buy some turbines at Värö, and formed the cooperative association Värövind ek för.

Håkan Svensson, head of the windpower department, formed a provisionary board and made an information brochure that was distributed to the utility's customers with their electric power bills. The 1,200 shares, corresponding to 1,000 kWh/year a piece, were sold out in about six months. All members were private households, not only from Varberg but from many parts of Sweden, even from Stockholm.

In 2003, the utility got a new board that was even more committed to creating a sustainable energy system. The new promotion system with certificates was introduced, and the new rules of the game became clear enough to make economic calculations.

The county council in Halland wanted to cooperate with Varberg Energi in a windpower project by offering a good piece of land with good wind resources. This resulted in a windpower plant with three 2-MW turbines at Munkagård. There was at this time also a long queue of people waiting for an opportunity to buy shares in the cooperative. Varbergsvind was offered to buy one of the new turbines, and this was then integrated in the cooperative. The number of shares increased from 1,200 to 7,000. This time, also a number of companies became members.

In the statutes, there is a rule that no member may own more than 10 percent of the shares, Håkan Svensson explains. Some companies also bought the maximum number of shares: 700 shares for a price of 6,000 SEK/share adds up to an investment of more than 4 million SEK. These companies were convinced that this would be a good business investment, Håkan Svensson concludes.

The cooperative association does not have any persons employed or any administration. On their bill from Varberg Energi, members pay 0.15 SEK + VAT, which gives a total of 0.19 SEK.[3] Compared to the market price, this will give each member a reduced cost of 500 SEK/share (1,000 kWh/year). Members who leave the association or want to reduce their number of shares will get the nominal value of the shares returned from Varberg Energi.

From a strict economic point of view, it would be more profitable for Varberg Energi to own the wind turbines, Håkan Svensson explains. However, by this local ownership, we get a better acceptance to build windpower plants. We engage our citizens and customers by letting them be active owners of power plants.

After deregulation, all citizens of Varberg were no longer automatically also customers of the local utility. However, as members of a cooperative, they also have to be customers of Varberg Energi. And those from areas outside the municipality who own shares in Varbergsvind also have to become customers. This is, of course, an advantage for the utility.

Varberg Energi has developed windpower plants for two decades and is now developing several new projects with a total of seventy-five to eighty wind turbines. Most of them will be sold to other actors, but the utility will keep some ten to fifteen of the turbines for themselves and for the Varberg windpower cooperative so that it can continue to expand.

More municipal utilities should develop windpower projects and promote windpower cooperatives, according to Håkan Svensson. In that way, the municipalities will be able to govern the development of a renewable energy system. The option with local ownership in windpower cooperatives should be part of the planning. This will make people in the vicinity engaged in windpower from start, rather than opposing it. And it is good business as well (Wizelius, 2010/2012a).

7.2.4 O2 El: National windpower cooperative

In just a few years, the company O2 El has grown to become one of the largest windpower cooperatives in Sweden. It has close links to the company O2, which was founded in 2005 and develops, builds, and operates windpower plants. For O2, the formation of cooperatives has been a successful business model. By the end of 2012, the O2 cooperative had some 4,000 members who together owned ten large wind turbines.

O2 El has two main driving forces. The first one is to create a renewable energy system, and windpower is the most cost efficient option right now. The second is to break up the ownership structure on the electric power market, explains Linda Magnusson, CEO at O2 El ek. för.

O2 El offers all to become part-owners: local inhabitants, municipalities, and companies. It offers local households to buy shares in the cooperative association, with preference according to the distance to the windpower plant.

To own shares in a cooperative association is a good model, according to Linda Magnusson. By buying shares that correspond to their own electric power consumption, citizens get a chance to do something to promote the environment and at the same time disconnect themselves from the price circus on the Nord Pool spot market. They get cheaper electricity and a predictable price.

The company O2 was founded in 2005 by two entrepreneurs with a background in the IT business. The vision of O2 is a green society with economic growth. At that time, there was consequently no "green" actor on the energy market. The company then also bought the windpower project developer *Vindkompaniet*, which had been in this business since 1991 and one of the most competent and experienced windpower development companies in Sweden, started by the persons who initiated and developed the very first windpower cooperatives on Gotland in the early 1990s.

People who want to buy shares in a wind cooperative register on the company Website and then pay their shares in advance. When a new project is in the pipeline and it is time to start to build a new windpower plant, people in the vicinity get information and offers to buy shares, by letters in the mailbox and advertisements in local newspapers. When the local inhabitants have registered for shares, people from other places in Sweden can get those that are left.

Since O2 has many projects in the pipeline in different parts of Sweden, no one has to wait very long before they get windpower electricity delivered. The costs for new shares is decided by the marginal cost (i.e., the cost for the last wind turbine built and connected), but as soon as it starts operating, the new members' shares become part of the portfolio of all wind turbines owned by the association and original members become part-owners in the new turbines added. With this model, there is no end of delivery and no problem when some old turbines are taken out of service after some twenty years.

Members are recommended to buy no more than around 80 percent of the estimated power consumption. The company has its own power trading company, O2 Vindel AB. Members have to use this company as power trader and buy the remaining power (i.e., the power that the wind turbines do not produce) from them.

This model with national windpower cooperatives, with wind turbines as well as members spread over Sweden, was first developed by the municipal utility Falkenberg Energi when it formed Sveriges Vindkraftkooperativ ek för in 1998. In 2012, SVEF has about 1,800 members (see Section 7.6.1).

O2 is a private company with strong financing from investment companies on the finance market. The company has high ambitions and builds a large share of new windpower plants in Sweden, and has the ambition, according to Linda Magnusson, to turn about 10 percent of them into cooperatives.

There is an enormous social power in cooperative ownership, Linda Magnusson explains. More than 500 persons attended the inauguration of our windfarm Hedboberget, where the church choir from neighboring Rättvik performed a newly composed windpower cantata.

O2 has also in cooperation with the bank Swedbank constructed a loan to make it easier for prospective members to finance their shares. With this arrangement, anyone interested in becoming a part-owner in a windpower plant will have the opportunity.

Locally owned power plants that produce electric power for the owners' own consumption will increase competition on the power market and should in the long run lower the consumer price for electricity on the market, according to Linda Magnusson.

7.2.5 Motives of cooperative members

There are several different motives to start windpower cooperatives. Windpower cooperatives started by engaged individuals, such as Christer Larsson in Lysekil (Section 7.2.1) have a personal interest in renewable energy technology and a political ambition to create such a system. By initiating and building wind turbines, they contribute to this. And to engage others with the same ideas and ambitions to become members of windpower cooperatives, they are able to finance such projects. The same motives were behind the start of the wind cooperative Kvarkenvinden in Umeå (Section 7.3.2). These were the main driving motives in the early 1990s, when there were no economic incentives and high risk.

These motives are probably still strong for many people today who buy shares and become members of cooperatives but no longer the only or the strongest one.

In the second (started in 1995) and third (started in 2000) cooperatives, some joined only for economic reasons, to make good business investments, according to Christer Larsson from Lysekil.

For Kvarkenvinden in Umeå, the motive to start this cooperative was idealistic, but today the motive to become a member is economic: members save significant amounts of money on their electric power bills.

Another interesting aspect is the changes of membership categories in the windpower cooperatives during this period. In the early 1990s, most members were private households. Today the majority of shares in many cooperatives are owned by small and medium-size companies, both private and public. Since these companies have to give the economic aspects priority, their main motives must be economic. This proves that the price for electric power from a cooperative is competitive (i.e., evaluated by the investors to be lower than on the Nord Pool market).

The project developer O2 has a policy to turn 10 percent of their wind turbines into cooperatives. For the developer, it will give stronger local support for their projects. The CEO of the national cooperative O2 El also mentions other motives: to give citizens an opportunity to promote the environment and disconnect themselves from the "price circus" of the Nord Pool market.

The motives of windpower cooperatives initiated by municipal utilities will be discussed in Section 7.6.

7.3 Windpower limited companies

There are many windpower limited companies that operate windpower plants in Sweden. Some of these are local, where a number of people from a local community or region have invested in a local windpower plant. Other limited companies are involved in project development and keep some of the turbines within the company.

A windpower limited company is by definition a business venture, which according to its statutes has to optimize the economic result for its shareholders. The motives of these companies is thus by definition economic. The individual shareholders could, of course, have other motives, as to benefit a sustainable society and so on, and have chosen to invest in a windpower company rather than in a power company that operates nuclear power plants or some other business. However, I have not hadinterviews with individual shareholders in windpower limited companies or cooperatives.

Windpower limited companies owned around 155 MW in 2003, and this figure increased to 1,775 MW in 2012. The share of wind limited companies increased from around 41 percent to 51 percent.

7.3.1 Ölandsvind AB

The company Ölandsvind AB was formed in 1991. It is based in Grönhögen on Öland, where there is a windpower plant with Vestas V27 wind turbines (see Figure 7.1), and most shareowners are local inhabitants from this area. Ölandsvind AB invested in one of the wind turbines.

Those who bought the shares at that time did not expect to earn a lot of money on the investment; it was more an opportunity to make a resolute concrete effort for the environment.

Over the years, the company has expanded, and today it owns three of the wind turbines in the Grönhögen windpower plant and has also invested 500,000 SEK in another windpower plant in Istad on Öland.

Figure 7.1 Windpower plant at Grönhögen, Öland, Sweden. Photo: Tore Wizelius

This company was once initiated by Olof Karlsson, who has been a driving spirit for renewables for decades and is chairman of the Swedish energy association's national organization, SERO. He has also been the CEO of Ölandsvind AB from the start and gets a remuneration of 2,000 SEK a month from the company, which is a symbolic sum (Hedman, 2007).

7.3.2 *Åsle Vind AB*

The windpower company Åsle Vind AB was formed in 2005 by fourteen entrepreneurs in the village Åsle in Falköping municipality. The company was initiated by the farmer Berndt Bolms, who at that time already had some experience of windpower. He had been a board member of the windpower production cooperative Slättens Vind since 2002.

Åsle Vind AB was formed to create a common project for a group of neighbors in Åsle, Berndt Bolms explains.

This company initially bought half a wind turbine in a windpower plant in Mungseröd on the west coast of Sweden. The windpower plant in Mungseröd consists of six 1.5-MW turbines. The different owners have formed a common company, Mungseröd Vindpark AB. Åsle Vind AB (which owns half a turbine) owns one-twelfth of the shares in this common so-called umbrella company. All earnings from the six turbines go to this company, and all expenses for operation and maintenance are paid by it. The earnings minus the costs for all turbines are then divided among the owners in proportion to their shares in the windpower plant.

In 2007, Åsle Vind AB invested in a local windpower plant in Skörstorp, which started to operate in February 2008 with three 2-MW wind turbines. The company owns a share in one of these; the rest are owned by other local limited companies, one windpower cooperative, and some private investors. This windpower plant is also operated by a common umbrella company, GSG Vind AB (Wizelius, 2009/2012).

7.3.3 Arise Windpower AB

The company Arise Windpower AB, based in Halmstad on the Swedish west coast, was founded in 2006 and was registered on NASDAQ OMX Stockholm stock exchange in 2010. According to the company's Website:

> The company's business concept is to manage the entire value chain as an integrated wind power company, from product development to selling green electricity produced by its own land-based wind turbines. Arise Windpower's goal is to produce around 2 TWh of environmentally-friendly electricity per year by 2014, equivalent to Sweden's total domestic wind power expansion at year-end 2008.
>
> (Arise Windpower AB, 2013, my translation)

By the end of 2012, Arise Windpower had forty-two employees, 139 MW windpower in operation or being operationalized, and another 102 MW under construction. More than 500 MW were in the permit phase. Chairman of the board is the former CEO of Volvo, Pehr Gyllenhammar, and one of the board members is Maud Olofsson, former minister of industry. The principal owners of the company are, besides the founders, public and private pension funds, insurance companies, the Norwegian state-owned power company Statkraft AS, and the windpower project developer Ownpower Projects AB.

7.3.4 Rabbalshede Kraft AB

The company Rabbalshede Kraft AB was founded in 2005 by two entrepreneurs from Rabbalshede on the west coast of Sweden. In 2012, Rabbalshede Kraft had twenty-four employees who worked on constructing onshore wind farms throughout Sweden. Rabbalshede Kraft AB plans and establishes onshore wind farms in Sweden to operate proprietarily or for sale to other companies.

So far, Rabbalshede Kraft has participated in the commissioning of forty-eight wind turbines, including twenty-six wind turbines in four wind farms operated by the company. Rabbalshede Kraft has been granted permits for another fifty-two wind turbines, of which fourteen are under construction. The project portfolio currently comprises about 450 wind turbines, distributed among some fifty windpower plants throughout Sweden.

According to the company's Website:

> Rabbalshede Kraft's overriding objective is to create value growth for the Company's shareholders—proprietarily, in partnership with other

> companies and through the divestment of entire farms and individual wind turbines—by contributing to the commissioning of 250 turbines by 2015.
> (Rabbalshede Kraft AB, 2013, my translation)

To guarantee the long-term capital requirements, the company is preparing an application for listing on the NASDAQ OMX Stockholm stock exchange.

7.3.5 Motives of windpower limited companies

These four windpower limited companies described in Sections 7.3.1–7.3.4 are of very different character. The first two are small local companies; the last two are public limited companies that are, or intend to be, listed on the main stock exchange in Sweden. However, all limited companies have, according to their statutes, the aim to optimize the profits for their shareholders. The two small local companies mention some other motives as well: Ölandsvind AB *to make a resolute concrete effort for the environment* (by financing renewable windpower) and Åsle Vind AB to *create a common project for a group of neighbors,* which is a social motive.

7.4 Other companies

Other companies are defined as companies in a line of business other than energy production. These companies can invest in wind turbines that deliver power to their factories or to produce electric power as a side business. Windpower ownership can also be a part of the company's environment policy: to create goodwill among its customers.

The first "other companies" invested in wind turbines already in the early 1990s. At that time, a grocery store and a shoe retailer bought their own wind turbines on Gotland. A fodder factory operated by the farmers' cooperative installed a wind turbine on its land in the harbor of Klintehamn and connected it directly to the factory. A few years later, a neighboring sawmill did the same.

These examples illustrate the two options that other companies can use, invest in, and own grid-connected wind turbines or connect such power plants directly to their factories.[4] Several large companies in Sweden have installed windpower plants, and more are in the pipeline.

Wind turbines connected directly to a factory, without taking the detour through the power grid managed by the grid operator, get this power at cost price. The turbines are just like any other machines in the factory. The power is produced at cost price and delivered as power (kWh). There is no electricity tax or VAT and no grid fees. However, if the power produced is measured properly, it will give the owner certificates that can be sold on the certificate market.

To make this option feasible, the factory should have production day and night, all year round. The maximum power of the turbines should be lower than the power of the factory.

The other option used is for other companies to build windpower plants connected to the power grid. Since around 2006, large industrial companies such as the forestry and paper pulp companies SCA and Holmen, the real estate company Wallenstam, and also the international furniture retailer IKEA have invested in windpower plants.

Some of the largest and most energy-intensive industries in Sweden founded a common windpower development company, VindInAB in 2006.

Other companies owned 54 MW in 2003, but this figure increased to 503 MW in 2012. The share of other companies increased from around 14 percent to 15 percent.

7.4.1 SCA Energy AB

SCA is one of the largest forestry and paper mill industries in Sweden. The company owns as much as 2.6 million hectares of forestland in the northern part of Sweden.

The daughter company, SCA Energy, develops renewable energy from forest resources and produces refined and unrefined forest-based biofuel. It also manages SCA's windpower assets, amongst them a number of windpower plants in northern Sweden in a joint venture with Statkraft AS,[5] Statkraft SCA Vind AB, which was founded in 2006.

Since then, this company has become one of the largest windpower developers in Sweden. By the end of 2012, four large windpower plants were in the building stage, and another three have permissions granted and will be built during the next few years.

All of them are very big windpower plants and built in the counties Jämtland and Västernorrland between the towns Sollefteå, Strömsund, Östersund, and Ånge, most of it on land owned by SCA.

The average size of these windpower plants is around fifty wind turbines, with 175 MW installed capacity producing almost 310 GWh/year each. The total of the four plants that are ready for operation is 210 wind turbines with a capacity of 915 MW, which will produce some 1.5 TWh a year.

To this, another 120 wind turbines with a capacity of 315 MW and a production of 615 GWh will be added in 2013–2014. Then Statkraft SCA Vind AB will have 330 wind turbines with a capacity of 1,230 MW in operation and produce more than 2 TWh annually (see Table 7.1).

This will be more than 1 percent of the power production in Sweden. Besides this, SCA Vind AB has more windpower projects in the pipeline.

This is an astonishing effort by an industrial company. Why is the company doing this?

According to the SCA Website:

> Sweden needs more renewable energy.
>
> It is a fact that the power consumption in Sweden increases faster than the power production. Sweden does not manage a cold day in January without importing electric power. And the power prices are rising.

Table 7.1 Statkraft SCA Vind AB windpower projects

Name	*Wind turbines*	*Capacity MW*	*GWh/y*
Building stage			
1. Stamåsen	50	115	250
2. Mörttjärnberget	37	100	210
3. Ögonfägnaden	33	100	290
4. Björkhöjden-Björkvattnet	90	600	800
Total in building stage	210	915	1,550
Permission granted			
5. Raftsjöhöjden	10	25	50
6. Bodhögarna	80	200	365
7. Storhögen	30	90	200
Total for 5–7	120	315	615
Total for 1–7	330	1,230	2,165

> SCA is an electric power intensive industry and thus needs to secure its power demand to stay competitive in the future. To do this with energy sources which do not affect the climate is self-evident. Statkraft is the second largest distributor of renewable energy in Europe and has large experience of windpower. To become a larger actor in Sweden is a way to continue the development of electric power production adapted to the environment.
>
> This joint effort by Statkraft and SCA is one of the largest industrial investments in Sweden: 16 billion SEK for windpower plants in forest areas in the regions of Västernorrland and Jämtland.
>
> (Statkraft SCA Vind AB, 2013, my translation)

From SCA's point of view, one of the motives is to get more value out of the forestland the company owns.

In 2005–2006, the company realized that the wind resources were very good in many of their forest areas. They began looking for partners and found Statkraft, which wanted to establish itself as a power producer in Sweden.

By this joint venture, Statkraft SCA Vind AB, Statkraft got access to land for large windpower plants, and as a part of this cooperation, SCA got a long-term power purchase agreement (PPA) to get electric power at a very competitive price (Westberg, 2013).

At the same time, SCA, with a 40-percent share of the windpower production, established itself as a power producer. All power is sold at spot prices to the Nord Pool market, and by being on both sides (i.e., both a power producer and a consumer), it works as a hedging of the cost for electric power for SCA.

Only 3 percent to 5 percent of the forestland is used for foundations and access roads, so the forest resources keep their value. The access roads can also be used for forestry; they add value to the area.

Today, there is no connection between the power produced by the windpower plants and the power used by SCA.

If it becomes necessary to use power from renewables to sell paper, we can arrange with origin marking; this can be arranged in the future, Åke Westberg, CEO of SCA Energy AB, explains (Westberg, 2013).

There are three motives for SCA to develop windpower. One is to get added value from the forestland owned by the company, the second is to hedge electric power prices (by long-term PPAs and addedpower production), and the third is to get the option to use renewables for the company's products.

7.4.2 Wallenstam

Wallenstam was founded as a construction company in 1944. By the end of the 1970s, it changed to become a real estate company. In 1984, the company was listed on the stock exchange and expanded in the following years from the Gothenburg region to Stockholm and other regions in the southern and middle parts of Sweden. The value increased from 500 million SEK to 3.5 billion SEK during the 1980s. Today, its real estate is concentrated in Gothenburg, Stockholm, and Helsingborg. After the turn of the century, Wallenstam added building construction to its activities.

In 2005, Wallenstam started to explore the possibilities to produce its own renewable energy. The first wind turbines were bought in 2006 and, since then, windpower development has continued at a fast pace. The target of Wallenstam is to have its own renewable energy production that covers the company's and its customers' demand.

The electric power Wallenstam produces is offered to their customers (i.e., tenants, and others). The power comes from wind turbines which produce the same amount as the tenants consume.

According to the company Website:

> Besides reducing emissions of greenhouse gases, our own produced power gives us a better control of the costs. This increased the security and protection for us and for you as a tenant—in line with our business philosophy.
>
> (Wallenstam, 2013)

Wallenstam also notes that:

> Our efforts with renewable energy have also had a positive impact on our chances to get land allocation agreements for new buildings, since our environmental ambitions are perceived positively and gives us a competitive advantage.
>
> (my translation)

By the end of 2012, Wallenstam had wind turbines installed at thirteen different sites, from single turbines to small windpower plants with up to ten wind turbines. The total number of turbines was forty-one, and the total installed capacity around 80 MW. Another 27 MW at two new sites were under construction and eight windpower projects under development. Wallenstam Naturenergi also has three hydropower plants.

Primarily, this is an environment goal. The target is to supply the company group and its customers with 100 percent renewable energy. Today it is no problem to make this a good business. Another motive is to secure the future power price; it is very expensive to secure this for ten to twenty years. Finally, it creates goodwill for the company, Bo Strandberg, CEO of Wallenstam Naturenergi AB, explains.

7.4.3 IKEA

IKEA is a Swedish furniture company with global presence; there are 287 IKEA warehouses in twenty-six countries (August 31, 2011). The company has set the target to use 100 percent renewables. According to the company Website:

> IKEA has made large investments in renewable energy moving towards our long-term target to use 100 percent renewable energy. More than half of the energy for our buildings is now delivered from renewables. Photovoltaic panels on 40 IKEA buildings in seven countries and 60 wind turbines produced 152 GWh "green" electricity this year (2011), which corresponds to about 12 percent of the electric power needed for all IKEA warehouses and distribution centers.
>
> (IKEA, 2012, my translation)

In Sweden, IKEA owns a windpower plant with nine wind turbines on Korpfjället in Dalecarlia since 2011. IKEA has ordered another windpower plant with thirty wind turbines at Glötesvålen in the region Härjedalen, which will be in operation in January 2015. With these two windpower plants, IKEA will lower its cost for electric power and will be self-sufficient with renewable energy for all its warehouses, distribution centers, and so on in Sweden. According to a press release from IKEA:

> Sustainability is a natural part of all IKEA does. It also gives us the opportunity to transfer the economic benefit of electric power from our own windpower plants into lower prices which will benefit our customers, Peter Agnefjäll, CEO for IKEA Svenska Försäljnings AB says.
>
> (O2, 2012, my translation)

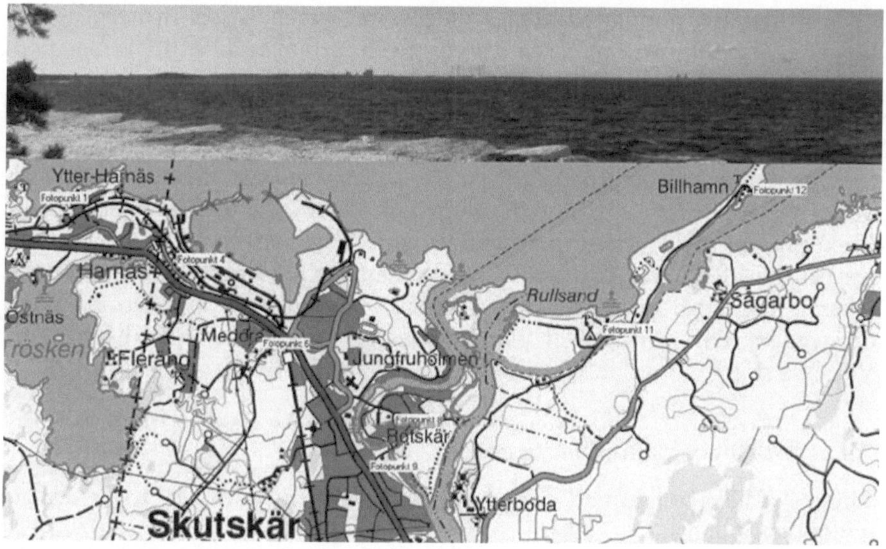

Figure 7.2 Wind turbines connected to a paper pulp factory in Skutskär. Five wind turbines have been installed in the industrial harbor, and the generated power is fed straight into the factories

7.4.4 VindIn AB

Some of the largest energy-intensive industrial companies in Sweden have formed their own windpower development company, VindIn AB. Among the owners are the forestry and paper pulp companies Stora Enso, Holmen, Billerud, and Korsnäs; the mining companies LKAB and Boliden; the cement manufacturer Cementa; and the oil company Preem, with large refineries at Lysekil on the west coast.

The ambition is to install windpower plants that produce 1 TWh/year within a five-year period.

VindIn has so far developed and installed a group of five wind turbines with a total capacity of 10 MW next to Stora Enso's factory in Skutskär, which have been on line since 2009. They produce around 30 GWh/year (see Figure 7.2).

In November 2012, the first stage of the windpower plant at Trattberget-Skalberget in the north of Sweden, with thirty wind turbines, was inaugurated. Within four years, it will be developed to a total of 150 wind turbines.

VindIn AB has several other large windpower projects in different stages of the development process. In Lysekil on the west coast of Sweden, one of the member companies, Preem, has a large oil refinery. VindIn has submitted an application to the county administration to build seven wind turbines on and next to Preem's industrial estate, in an area designated for industrial use. These turbines will produce 55 to 65 GWh/year.

In other less populated areas, VindIn has much larger projects in the pipeline. A windpower plant with sixty-seven turbines and a total capacity of 134 to 268 MW is developed at Suorva in the fjells in the north of Sweden. Another windpower plant with thirty to fifty wind turbines (60–200 MW) is planned at Sävar north of Umeå on the coast of the Baltic Sea in northern Sweden. A third large project with thirty-seven wind turbines and a capacity of up to 140 MW is being developed in Vedbo in the forested county Jönköping in the southern part of Sweden. VindIn AB has also acquired 10 per cent of Blekinge Offshore AB, which has applied for permission to build a very large offshore windpower plant in Hanöbukten, on the south coast of Sweden. This windpower plant would produce some 8 TWh/year, and VindIn's share would be 800 GWh/year.

7.4.5 *Other companies' motives*

These "other companies" have several different motives to invest in windpower plants.

The forestry company SCA mentions three: to get added value from the forestland owned by the company, to hedge electric power prices (by long-term PPA), and to secure the option to use renewables for the company's products, if this will be demanded by their customers.

For Wallenstam AB, it is an environment goal to supply the company group and customers with 100 percent renewable energy. Another motive is to secure the future power price, and finally it creates goodwill for the company.

IKEA will lower its cost for electric power and will be self-sufficient with renewable energy for all its warehouses and distribution centers. Sustainability is also a part of the company strategy, and investment in renewables will strengthen its trade mark.

VindIn AB has the task to develop cost-efficient windpower plants for its owners, and it is the motives of the owners that are interesting.

The main motive of other companies is to secure stable long-term power prices by producing it with windpower plants that have fairly predictable costs. This is a long-term strategic interest that motivates these very large investments in new windpower plants by these companies. Some have additional motives, as to strengthen their trade mark (IKEA, Wallenstam) and utilize land areas for new purposes (SCA). However, business companies, especially those who are registered on the stock exchange, have to make investments that, based on current knowledge, should be profitable.

7.5 Power companies

In Sweden, the three power companies Vattenfall, E.ON, and Fortum are dominating the electric power market. Vattenfall is the Swedish state-owned

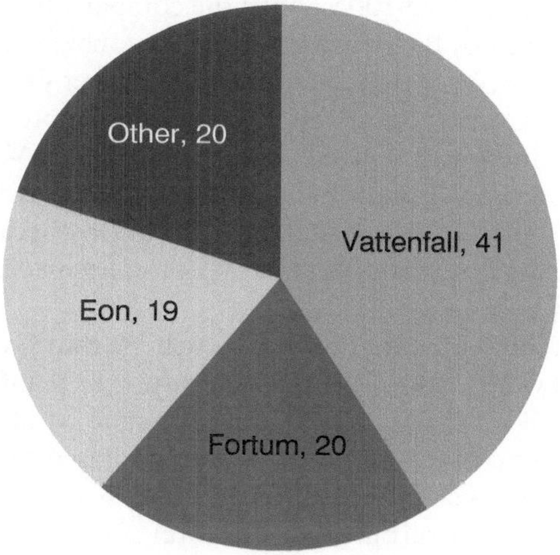

Figure 7.3 Electric power production in Sweden, 2011, percent (Energimyndigheten, 2012)

power company. Fortum is the Finnish state-owned power company, and E.ON is a private limited company based in Germany. Also the Norwegian state-owned power company Statkraft is active on the Swedish market.

In 2007, Vattenfall produced 62 TWh (44 percent), E.ON 31 TWh (22 percent), and Fortum 25TWh (18 percent). The remaining production came from Statkraft (2 TWh) and other smaller producers (20 TWh). These power companies use all kinds of power plants: hydro, nuclear, coal, gas, and renewables. They also act on an international market and export or import power to the Nordic power market (see Figure 7.3).

7.5.1 Vattenfall

Vattenfall is the fifth largest energy producer in Europe. Vattenfall has been involved in windpower development since the end of the 1970s; and in 1983, the first large 2-MW prototype was installed at Näsudden, Gotland, as a part of a state-financed research and development program.

During the 1990s, Vattenfall invested in several other wind turbines at different sites in Sweden as part of its research and development program and to learn to manage and evaluate this new technology.

In 2006–2007, Vattenfall built the so-far largest offshore windpower plant in Sweden, with forty-eight wind turbines with a total capacity of 110 MW, at Lillgrund between Malmö and Copenhagen in Öresund.

In 2012, Vattenfall owned and operated about 900 wind turbines in Sweden, Denmark, Germany, the Netherlands, and the United Kingdom. They produced some 3.7 TWh annually. Among these are the offshore windfarm Horns Rev in Denmark, built in 2002 by the Danish power company Elsam and owned by Vattenfall since 2005, when the company purchased the Danish power company as part of the company's international expansion.

In November 2008, Vattenfall acquired the Thanet Offshore Wind Farm project in England. Construction was completed in September 2010. There are 100 wind turbines with a total capacity of 300 MW. It was then the largest offshore windpower plant in the world. The Kentish Flats wind farm off Herne Bay/Whitstable, Kent, England is also owned by Vattenfall. In early 2012, Ormonde Offshore Wind Farm in the Irish Sea, with thirty wind turbines of 5 MW, became operational.

In Sweden, Vattenfall operated some 120 wind turbines with a total annual production of 730 GWh. Östra Herrestad windpower plant in Simrishamn in the south of Sweden was built in 2010, and the nine wind turbines were put in operation at the end of the year. Vattenfall purchased this project in 2008.

The largest windpower plant on land is Stor-Rotliden, in the north of Sweden, with forty wind turbines with a total nominal power of 78 MW, which produce some 240 GWh a year. This windpower plant was built in 2011.

In 2006, Vattenfall began to consider windpower as a competitive power source and started to make large investments in windpower plants. Most of these, around 90 percent, were made in other countries with fixed price systems, such as the United Kingdom, Denmark, and the Netherlands.

Vattenfall is owned by the Swedish state; the board members are appointed by the government, which also formulated its policy. In 2006, the company's CEO Lars G. Josefsson declared that Vattenfall had the ambition to develop new renewable energy sources in Sweden that can produce 10 TWh a year in 2016. By 2012, the power production of Vattenfall had increased only from 32.2 to 32.7 TWh.

The share of investments in renewables has decreased, from 29 percent in 2006 to 18 percent in 2011 (total investments on all markets, even outside Sweden). During the same period, Vattenfall investments in fossil fuel–fired power plants increased from 31 percent to 52 percent. In real terms, the investments in renewables increased; the investment volume in 2011 was three times higher than in 2006 (Ny Teknik, 2012).

7.5.2 E.ON

Sydkraft AB, a power company owned by a number of municipalities in southern Sweden, built and operated the second large prototype wind turbine at Maglarp from 1982 until it was closed and dismantled in 1993. Sydkraft AB was bought by E.ON in 2001, but the name was changed from Sydkraft to E.ON in 2004.

E.ON Vind develops windpower in Sweden, Denmark, Norway, and Finland. In 2012, the company had a total installed capacity in Scandinavia of 287 MW. In 2011 and 2012, the company invested around 2 billion SEK in windpower plants in Sweden.

In 2010, one of the world's largest offshore windpower plants, with ninety wind turbines, started to operate outside Rødby in Denmark. In Sweden, E.ON Vind operates a number of windpower plants, most of them with three to ten wind turbines. The oldest one is from 1996 with twelve turbines on Gipsön, an artificial island outside Landskrona. One is from 2001 in Malmö harbor and all the rest from 2011–2012. A medium-size offshore windpower plant outside Öland, with sixteen wind turbines and 48-MW capacity, will be taken into operation in 2013.

By the end of 2012, E.ON owned 3.5 percent of the installed windpower capacity in Sweden, which is a marginal share for a company that produces more than 20 percent of the electric power in Sweden.

7.5.3 Fortum

Fortum did not have any significant activity in windpower development before 2010, when it formed the joint venture Blaiken Vind AB with Skellefteå Kraft (see Section 7.6.7). Fortum owned minority shares in some single wind turbines in Finland before that but has now a few other projects under development, one in Sweden and some others in Finland. Hydro power plants have a large share of Fortum's power production, 29 percent, but only 3 percent are "new" renewables, biomass, windpower, and so on; 34 percent comes from nuclear power plants, and the remaining power comes from fossil fuels, according to Fortum's sustainability policy formulated in the brochure "Brighter times ahead" (Fortum, Brighter Times Ahead, 2012).

According to the company Website, Fortum's policy and motives are:

> The aim is to reduce emission from fossil fuel fired plants and make them more efficient. Fortum believes that there will be *a gradual transition from traditional energy production towards a Solar Economy.*
>
> (Fortum, 2013)

There are no specific motives stated for investments in windpower or other renewables.

7.5.4 Statkraft

Statkraft AS is owned by the Norwegian state. In 1997, Statkraft decided to develop windpower plants, and the first one, on the island Smøla, was in operation in 2002. This is a very large windpower plant with sixty-eight wind turbines, a capacity of 150 MW, and an annual power production of 450 GWh. In 2012, Statkraft owned and operated three windpower plants in Norway and had some more projects under development.

Statkraft also develops windpower in the United Kingdom, on land and offshore. In England, Statkraft and the state-owned Norwegian oil company Statoil have developed the offshore windpower plant Sheringham Shoal, with eighty-eight wind turbines and a total capacity of 317 MW. It will produce 1.1 TWh a year. Statkraft has set the target to develop 1,100 MW windpower on land and 150 MW offshore before 2015 (Statkraft, 2013).

In Sweden, Statkraft got involved in windpower in 2006 when it formed the joint venture Statkraft SCA Vind AB, with the large Swedish forestry and paper pulp company SCA (see Section 7.4.1). In the southern part of Sweden, Statkraft cooperates with Södra Skogsägarna,[6] a farmer-owned forestry and paper pulp company (Södra, 2013). Statkraft is developing forty windpower projects in cooperation with these two partners.

The first windpower plant in Sweden with four wind turbines was on line in 2011, next to the large paper pulp factory in Mönsterås. It is operated by Statkraft Södra Vindkraft AB and produces around 25 GWh/year.

The company Statkraft SCA Vind AB is developing six large windpower plants in the counties Västernorrland and Jämtland, which is one of the largest industrial investments in Sweden ever. The project covers 122 km^2 forestland and 330 wind turbines. When all of these are online, they will produce 3 TWh per year (Statkraft SCA Vind AB, 2013).

Statkraft and Södra made an agreement in October 2009 to develop about 350 wind turbines in southern Sweden, which will produce around 1.7 TWh per year. Projects are in the development stage at twenty-two different sites.

Statkraft sees a large potential for further development of windpower on land in Norway, Sweden, and the United Kingdom. In the United Kingdom, there is also a good potential for offshore windpower, and the British government has set a target to develop 25,000 MW offshore. Statkraft is applying for permissions to build both on land and offshore (Statkraft, 2013). Statkraft will within a few years be one of the largest windpower producers in Sweden.

7.5.5 Motives of power companies

These power companies belong to the dominant energy system (DES) and act as transnational companies on an international market. Their motives are to defend and, if possible, expand their market shares and to deliver a good surplus to their owners.

Vattenfall and Statkraft have made big investments in offshore windpower in the United Kingdom. The economic conditions for those investments are very favorable, giving a secure long-term power price, much higher than could be expected in Sweden. The government has given Vattenfall a mandate to develop renewables in Sweden. According to the articles of association, Vattenfall will be a company that leads the development of an ecologically sustainable energy production (Vattenfall AB, 2012).

Vattenfall still makes its largest investments in countries outside Sweden and more in fossil fuel than in renewables. Vattenfall has motives (the government mandate) but obviously lacks the motivation to develop windpower in Sweden.

Norwegian Statkraft has started to make large investments in windpower plants in Sweden. The motive for this is to get a foothold in the Swedish electric power market. Fortum's motives to invest in the windpower plant at Blaiken, in a joint venture with Skellefteå Kraft AB, is motivated by business interests and evaluated as a profitable investment.

E.ON is a private power company based in Germany. The other transnational companies on the Swedish power market—Vattenfall, Fortum, Statkraft, and Dong Energy—are public state-owned companies that get their directives from their governments. The politicians are thus responsible for the policy of these companies.

7.6 Municipalities and municipal utilities

Already in the early 1990s, some municipal utilities, for example Gothenburg, Falkenberg, Varberg, and Örebro, developed their own windpower plants. In 2012, many municipalities, municipal utilities, and also companies have their own windpower plants. Some of these municipal utilities have played an important role for the development of windpower in Sweden through investments while windpower still was a new and marginal power source, with some risk.

Also with the development of ownership models, they have played a creative role. The first windpower cooperative was developed by the municipal utility in Gothenburg, Göteborg Energi. The utility developed a couple of wind turbines in the harbor, developed a legal framework for a consumer cooperative, and marketed and sold shares to their citizens/customers in 1993. In those days, there was still a monopoly: all consumers in Gothenburg were also the customers of the municipal utility (Wizelius, 1993).

Since 2005, the interest of municipalities in windpower has increased significantly. Some have bought shares in wind cooperatives, such as the municipality Robertsfors, which in 2008 bought shares in Kvarkenvinden when the coop had developed a new wind farm in the municipality that covered the consumption in municipal buildings (schools, offices, etc.) and the municipal housing company Robertsfors bostäder (Wizelius, 2010/2012a).

The municipal utility Umeå Energi, which had played a very creative role in the development of the net-account model for Kvarkenvinden cooperative, developed several windpower plants along the coast of northern Sweden (Wizelius, 2010/2012b).

Utilities owned around 22 MW in 2003, and this figure increased to 450 MW in 2012; the share of utilities increased from around 6 percent to 13 percent.

7.6.1 Falkenberg Energi

The municipality Falkenberg on the west coast of Sweden has been at the forefront of the development of renewables for several decades. The politicians in this municipality have had a strong interest in environmental issues, and this has attracted civil servants with the same interests.

The planning architect Johan Risholm and the former CEO of Falkenberg Energi, Jan Åke Jacobsson, both moved to Falkenberg due to its ambitious energy policy. The first grid-connected wind turbine in Sweden was built in Falkenberg in 1983.

When he started to work as a planning architect in 1990, there was a plan to develop the harbor, Johan Risholm recalls. Falkenberg made a plan for windpower development in the harbor area, and this was the first municipal windpower plant in Sweden. Six wind turbines were then installed in the harbor area.

A few years later, a new highway (E6) was built just outside the city Falkenberg. The municipality then bought large areas of land for this development. Since the highway would pass outside of the city, the car drivers would not notice Falkenberg.

The politicians decided to develop a windpower plant just next to the highway, to make Falkenberg and its ambitions to develop a sustainable power system visible. This was the first task of Jan Åke Jacobsson, the new CEO of the municipal utility Falkenberg Energi.

The windpower plant was called the Falkenberg Gate, and according to the political decision, at least one of these wind turbines should be owned by the citizens of Falkenberg via a windpower cooperative. The windpower plant started to operate in 1998, and one of the turbines was owned by the cooperative Falkenbergs vindkraft ek för.

Windpower got hundreds of enthusiastic ambassadors, Jan Åke Jacobsson recalls. No one was opposing windpower.

Ventosum AB manages the windpower plant; the ownership structure is described in Section 5.8.1. In 2004, another four wind turbines were installed on the other side of the highway, forming a real gate (see Figure 7.4).

During this time, the wind turbines in the harbor were worn out, and the municipality decided to make a generation shift. In 2006, a repowering was made, and the old turbines were replaced by five new large 2.3 MW turbines, all owned by the municipality.

This proved to be a very good business idea, Jan Åke Jacobsson explains. They are visible from many parts of Falkenberg, but very few have raised any objections.

It is very valuable to have a municipal utility, Johan Risholm argues. A company owned by the municipality can be used as a tool to realize political ideas and ambitions. The utility has been allowed to take risks by testing new technology.

It is important to give citizens the opportunity to participate in the projects, according to Jan Åke Jacobsson. If a large company enters an area, develops a windpower plant, and then disappears, it is hard to get local support for the project. It is also important to argue and explain why

Figure 7.4 The Falkenberg Gate. The siting of these wind turbines was governed by the politicians to display its profile as a future-oriented municipality. Photo: Falkenberg municipality

it is necessary to develop windpower. We are not doing this to tease people but to solve environmental problems and stop climate change.

In the county Halland, several neighboring municipalities, such as Varberg (see Section 7.2.3), have also been very active in this field. The southern neighbor Halmstad has, however, not much windpower installed, although the wind resources are the same. This shows that municipal policy matters.

7.6.2 Dala Vind AB

The county Dalecarlia, a region characterized by forested mountains in the middle of Sweden, was a late starter in the windpower business. In 2008, however, there was a windpower boom in the county. In January, there were nineteen wind turbines operating; by the end of the year, sixty wind turbines were installed. In 2012, the power grid was full. New power lines have to be built before new projects can be developed.

The company Dala Vind AB has played a crucial role for this fast development of windpower in Dalecarlia. Many municipalities, municipal companies, and private citizens have invested in these new windpower plants.

Dala Vind AB was formed in 2006 by *Dala Kraft AB* (Dalecalia Power Ltd Co.) and the municipal utility in Malung, *Malungs elverk*. There are eighteen part-owners in this company (see Table 7.2).

Table 7.2 Dala Vind AB: owners

Name of part-owner	*Type of company/organization*
Dala Kraft AB	PPP Municipal (several municipalities) and private companies
Malungs Elverk AB	Municipal power company
AB Borlänge Energi	Municipal power company
Falu Energi & Vatten	Municipal power company
Envikens Elkraft ek för	Cooperative association (hydropower&grid)
Hedemora Energi AB	Municipal power company
Dala Energi AB	Grid operator (owned by several municipalities)
Smedjebackens Energi AB	Municipal power company
Prästlönetillgångarna i Västerås stift	Land owner, Västerås diocese & parishes
Lima besparingsskog	Forest owners association*
Transtrands besparingsskog	Forest owners association
Älvdalens besparingsskog	Forest owners association (2,300 members)
Särna-Idre besparingsskog	Forest owners association (1,100 members)
Hamra Kraft AB	Local power company owned by forest owners
Skistar AB	Private company—skiing business
Private investors (2)	Private

* Real estate commune for forest management

The purpose of Dala Vind AB is to develop windpower plants for the regional cooperative Dala Vindkraft ek för and for the owners of the company.

Dala Vind AB was started to make it possible for us to develop our windpower plants ourselves, Jan Högberg, who initiated this company, explains. If we manage the process ourselves, we get closer to the actual cost price. It is cheaper than to buy turnkey wind turbines from other development companies (Wizelius, 2010/2012b).

The origin of this company is an idea from the local inhabitants in the village Äppelbo. They asked a project developer to check the winds on a hill next to the village, where inhabitants knew that there was a lot of wind. The wind measurements proved that the winds where excellent, although the site was in a forested area. The villagers then asked the CEO of Malungs elverk, who lived in the same village, to assist them, and Malungs elverk and Dala Kraft AB helped them to finance the project and organize the windpower cooperative Äppelbo Vind ek för.

In 2006, Dala Kraft together with some other municipal power companies in the region formed the regional cooperative Dala Vindkraft ek för. They copied the model used in Falkenberg (described in Section 5.4.4). At the beginning of 2012, the cooperative Dala Vindkraft operates five wind turbines and has 2,100 members. There are also several independent wind cooperatives, which are administrated and managed by Dala Vind AB.

These two companies, Dala Kraft AB and Malungs elverk, then started the regional windpower development company Dala Vind AB. This company now owns five wind turbines and has four projects in the pipeline, with a capacity of more than 100 MW, waiting for permissions.

The main factors behind the success of Dala Vind AB are that the part-owners are a mix of landowners, grid operators, and power companies anchored to the territory, according to Jan Högberg.

They realized at an early stage that it was important to give local inhabitants the opportunity to become part-owners by becoming members of a windpower cooperative association. It makes the neighbors engaged and makes it much easier to get permissions to build windpower plants.

In Dalecarlia, most local inhabitants are in favor of windpower, according to Jan Högberg. The municipalities have also been rather positive, with a few exceptions. The opposition has come from civil servants at municipalities and the county administration. When we have public consultation meetings, we are met by applause, which is quite rare in other parts of Sweden (Wizelius, 2010/2012b).

7.6.3 Rättvik

In the municipality Rättvik in Dalecarlia, the municipality itself has invested in windpower. In 2006, the development company O2 Vindkompaniet got in touch with the municipality to get permission to develop a windpower plant on Hedbodberget, in this municipality. They told the developer that the municipality wanted to buy and own some of these wind turbines.

When big business comes to Rättvik to invest hundreds of millions in the middle of the forest, it must be possible to earn some money. So we wanted to share this profit, Inge Östlund, commissioner in Rättvik municipality, explains (Wizelius, 2010/2012b).

The developer O2 Vindkompaniet was not very interested to let the municipality buy one of the turbines. They suggested that the municipality should buy shares in the company's cooperative instead. However, the municipality insisted and finally got the opportunity to buy two of the nine turbines in this windpower plant. The municipality then asked Dala Kraft AB to establish a local cooperative, Rättvik vind ek för, which would get one of the two turbines.

This arrangement with local ownership was necessary to create local acceptance for the project, Inge Östlund explains. The developer was satisfied as well and will include offers for municipal ownership in their business model. The municipality has also reserved another wind turbine in the second stage of the project. Then Rättvik will get all its electricity from local windpower.

The municipality has also worked out guidelines for local benefits from windpower. If a developer is not prepared to pay 0.5 percent of the gross income to the municipality, we are not interested to let them build any windpower plants here, Inge Östlund explains. These funds for local benefits[8] are to be used for local development, and those with good projects can apply for grants from this fund once a year.

7.6.4 *Kalmar Energi*

The municipal utility in Kalmar, a town on the east coast of Sweden, bought its first windpower plant in 1996. It was a small group of wind turbines built on a pier next to a cement factory on the island Öland.

Kalmar Energi is a public–private partnership: Half of the company is owned by the municipality, the other half by the transnational power company E.ON. However, the municipality has the majority vote on the company board.

The first wind farm was quite small but had a symbolic value for Kalmar Energi, Staffan Larsson, head of the marketing division, explains. In 2006, a new larger wind turbine was developed close by the first wind farm, and Kalmar Energi purchased it together with two new wind turbines at another site on Öland. Kalmar Energi then offered its customers to become part-owners in these wind turbines by forming the windpower cooperative Kalmarsund Vind.

The idea behind this was to create long-term relationship with our customers, Staffan Larsson explains. It gives a strong local connection and a good environmental company profile. The wind coop has a contract with Kalmar Energi to manage the power balance, to administer the money in their invoice system, so that the power from the wind turbines can easily be subtracted from their bills. Kalmar Energi has used the same model as SVEF in Falkenberg (see Section 7.6.1).

Kalmar Energi manages the administration of other cooperatives as well. These have asked for tenders, and Kalmar Energi gave the best offer.

This is good business for Kalmar Energi. The company has a good invoice system, qualified staff, and good routines to manage this. The members of these cooperatives also will become customers of Kalmar Energi.

In 2008, Kalmar Energi tried to create a new windpower cooperative, Kalmarsund Vind II, but had to cancel this project. Initially, the sale of shares was very good, but the sales died completely when the new tax on cooperative ownership[9] was introduced, Staffan Larsson explains (Wizelius, 2010/2012a).

7.6.5 *Piteå*

The project developer Svevind AB has started to build one of the largest windpower plants in the world, with 1,101 multimegawatt turbines in Markbygden in northern Sweden. The project area is part of Piteå municipality. When all wind turbines are installed, this windpower plant will produce as much power as a nuclear power plant: 8 to 12 TWh a year.

In 2012, there were twelve wind turbines operating, testing the site to find the best building methods, configuration, hub heights, and so on. The project will run for many years, and if everything goes according to plan, the whole windpower plant can be ready for operation in 2020 (Svevind, 2013).

The municipality has so far not opted to become part-owner in this windfarm. Instead, the focus has been on getting the developer and supplier to establish factories in Piteå. Another part of the strategy is to prepare local companies to do as much of the work as possible on access roads, foundations, and technical parts. Also education and recruitment of service staff is part of the municipal strategy (Wizelius, 2010/2012b).

Piteå municipality has appointed a head for development of industry and energy, Erik Persson, to prepare local companies and negotiate with the developer and turbine manufacturer on building factories in Piteå.

This will give indirect benefits for the municipality, increased tax revenues, and higher employment rates.

So far, however, the municipality gets only increased costs, Erik Persson concludes. The municipality could, of course, become owner of some wind turbines, and this opportunity is now considered by the municipality.

7.6.6 *Umeå Energi*

The municipal utility Umeå Energi in northern Sweden has in just a year acquired seventeen wind turbines and invested 400 million SEK. The company has suddenly become a big actor in the windpower business in the county Västerbotten.

The turnaround happened in 2007, when the board decided to get on the train, explains Lasse Johansson, responsible for windpower at Umeå Energi, but also one of the initiators of the windpower cooperative Kvarkenvinden in 1998 (see Section 7.3.2).

The policy of Umeå Energi, defined by the local politicians on the board, is to use power only from renewable energy sources. Besides the windpower plants, the utility also has built a large CHP plant fired with biomass.

The motive of the municipal utility was political, but it was not realized until windpower became economical.

7.6.7 *Skellefteå Kraft*

In Skellefteå, some 135 km north of Umeå, the municipality owns Skellefteå Kraft AB. It is the fifth largest power company in Sweden, with several large hydro power plants, and generates large incomes for the municipality each year.

Skellefteå Kraft has 650 employees who manage power production and distribution as well as trade with heat and electric power. The company produces some 3.8 TWh a normal year (71 percent from hydropower, 4.4 percent from nuclear power, 3 percent windpower in 2011) (Skellefteå Kraft, 2013).

The company bought its first windpower plant, on the coast, with three 600-kW turbines in 2003. The second project was much larger: ten 3-MW turbines in the mountains, on line since 2010. The same year, another windpower plant with eight turbines at Storliden was installed and, in 2011, one more at Jokkmokksliden with ten wind turbines.

In two years, Skellefteå Kraft became one of the biggest windpower operators in Sweden. And there is another much bigger windpower plant in the pipeline. In a joint venture with the Finnish power company Fortum, Blaiken Vind AB, a project with 100 wind turbines at Blaiken in Västerbotten county has been developed. It will be built in three stages. The first stage has started, and by the end of 2012, twelve wind turbines were on line. In 2015, all 100 will be in operation.

Skellefteå Kraft AB offers up to five-year fixed prices for electricity from windpower to private customers (Skellefteå Kraft, 2013).

According to its policy, Skellefteå Kraft AB shall generate both business and environmental benefits by developing and introducing renewable energy sources and sustainable energy solutions.

7.6.8 *Jämtkraft*

Jämtkraft is a power company owned by the three municipalities Östersund, Åre, and Krokom in the county Jämtland. According to the company website:

> Jämtkraft is a different kind of power company. We produce clean energy with renewable energy sources. We do not do it to make money. We do it to serve our customers. This has been decided by our owners, the municipalities Östersund, Åre and Krokom.
>
> (Jämtkraft, 2012, my translation)

The company operates seventeen hydropower plants and has had a wind turbine in operation since 2002. In 2007 and until 2012, more windpower plants were built, one with five, one with ten, and one with thirty wind turbines.

In 2008, Jämtkraft formed a joint venture, JP Vind AB, with the local private company Persson Invest, which operates saw mills and also owns large areas of land. This company now has three large windpower plants in the pipeline, with more than 150 wind turbines, waiting to get the necessary permissions (JP Vind, 2012).

7.6.9 *Falköping*

Falköping is a rather small municipality dominated by farming in an area with very good agricultural land. It has around 30,000 inhabitants and an area of 1,000 km^2. There are forty wind turbines, single turbines at farms or small clusters, with a nominal power of 43 MW (Energimyndigheten,

2013B). Most of these wind turbines are owned by farmers, wind co-ops, or local windpower limited companies (see Section 7.1.2).

In December 2009, the municipality adopted a comprehensive plan for windpower. At the same time, the local politicians decided that the municipality would invest 70 million SEK to make the municipality self-sufficient with renewable electric power and to promote local ownership of windpower plants.

According to municipal commissioner Ulf Eriksson, the purpose of this policy is to promote windpower development in the municipality, and that a large share of it will be locally owned to benefit the local economy.

In a first stage, the municipality will acquire wind turbines to produce electric power needed for street lightning, schools, senior homes, offices, and premises owned by the municipality. To produce the 25 GWh/year needed for this, it takes about 10 MW windpower.

The municipality has decided to invest 70 million SEK. The investments will be made in joint ventures with other local actors. The municipality invests only if at least 50 percent of the other investors are private persons, farmers, or companies registered in the municipality. Then the municipality will invest a maximum of 15 percent, with money directly from the municipality's cash box.

So far, the municipality has invested in two windpower plants where the ownership is shared with local farmers. By 2014, all 70 million SEK have to be invested.

The local utility Falköping Energi was sold to another utility, Göteborg Energi, in the year 2000. This company also invests in windpower within the municipality. There is also a municipal housing company.

When the target for the municipality is reached in 2014, the housing company can continue with investments in windpower, according to Ulf Eriksson, who also is a member of the board in this company (Wizelius, 2010/2012b).

7.6.10 Karlstad and Hammarö

In May 2010, Vindpark Vänern, with ten 3-MW turbines, was formally inaugurated. This was the first freshwater offshore wind farm in the world, with the turbines installed on cost-efficient rock foundations, also a first in world technology. Five of the wind turbines were owned by municipal companies, two by cooperatives, and three by a private local investor.

Two municipalities, Karlstad and Hammarö, are part-owners in this windpower plant. The municipal utility Karlstad Energi AB owns two of the wind turbines, and the municipal housing company Karlstad Bostäder one. In the neighboring municipality Hammarö, the utility Hammarö Energi AB owns one and the housing company Hammarö Bostäder also one.

The project was initiated by a local NGO, Värmlands Energi och Vindkraftförening (Värmland Energy and Windpower Association; VEV), which had started the windpower cooperative *Hammarö Vind* in 1995. It took more than ten years to develop Vindpark Vänern. The municipalities played a crucial role to make it possible to realize this visionary project. To finance the project, a consortium was formed by Karlstad and Hammarö municipalities, some cooperative associations, and private investors. The idea was that the wind farm should be locally owned.

Public support was very strong, since most of the wind turbines would be owned by the municipalities where people lived, and households could become members of windpower cooperatives.

The members of VEV had a strong commitment to create a renewable energy system. This was the initial motive and driver for this project. The municipality has also some very good motives. There are political targets, but there are not only environmental but economic reasons for municipal companies to own their own power plants, according to Åke Pettersson Frykberg, chairman of Karlstad Energi AB (Wizelius, 2010/2012b).

7.6.11 Telge Energi

Telge Energi was established in 1994 during the deregulation of the electric power market and started as a power trading company in 1997. The company is owned by Södertälje municipality. Telge Energi sells only renewable hydro- and windpower to its customers: households, and enterprises all over Sweden. The policy of Telge Energi is to increase competition in the power market, to have a good environment profile, and to let the customers decide how the power is produced (Sundelius, 2013).

Telge Energi has run several marketing campaigns for windpower, by brochures and TV commercials, trying to get new customers and criticizing the dirty power from the larger power companies, such as Vattenfall, E.ON, and Fortum (Telge Energi, 2013).

Telge Energi buys wind-generated electricity on the market from windpower producers and distributes it to its customers. In 2011, Telge Energi formed the windpower cooperative Telge Energi Vindförening and offered its customers to buy shares. Some 2,000 customers are shareowners (2012), and Telge Energi Vindförening now owns two wind turbines.

While other cooperatives offer shares of 1,000 kWh/year, Telge offers shares of only 100 kWh/year for a price of 750 SEK. This is affordable for all customers and also a good option for small households with a low power consumption. The price for electricity will then be 19 öre/kWh, which is cost price. The share can either be paid up front or during a period of thirty-six or sixty months. The limit for shares is set to 80 percent of the estimated power consumption. Shareholders have to be, or become, customers of Telge Energi. All of this can be done very easily on the company Website (Telge Energi, 2013).

The arguments that Telge Energi uses to convince customers to buy shares are that they will escape from the electric power market, have predictable and lower power prices, contribute to the development of windpower in Sweden, and make coal-fired power plants disappear from the Nordic power market (Telge Energi, 2013).

In 2009, the company introduced the product "Windpower" and started to sell origin-marked electric power 100 percent from wind. The company's motives to sell only clean power from renewables is not based on any political directives from the owners. It is the business idea of the company, formulated and implemented by the management.

This is a value-governed company, Stefan Sundelius, who is responsible for windpower at Telge Energi, explains:

> All employees can be proud of their work. We work to create a better world. We increase the pace of the transition to renewables, by developing good products for our customers. To sell shares in windpower plants is a win-win offer. We get satisfied customers, and they get a stable and low electric power price, without any risk.
>
> (Sundelius, 2013)

7.6.12 Bixia Energi

Bixia AB is owned by ten different municipal companies or utilities and grid cooperatives[10] (see Table 7.3). It was established in 1997 in response to

Table 7.3 Bixia Energi AB owner companies

Name	Ownership, power production
Borgholm Energi	Municipal utility, local grid operator, owns windpower plants
Mjölby-Svartådalen Energi	Municipal utility, local grid operator, owns hydro- and windpower plants
Oxelö Energi	Municipal utility, local grid operator
Tekniska Verken i Linköping	Municipal utility, local grid operator, owns hydro- and windpower plants
Växjö Energi	Municipal utility, local grid operator, owns biomass CHP plant
Nässjö Affärsverk	Municipal utility, local grid operator, owns biomass CHP plants
Södra Hallands Kraft	Cooperative association. Local grid operator, owns windpower plants and biogas plant
Bjäre Kraft	Cooperative association. Local grid operator, owns windpower plants
Sandviken Energi	Municipal utility. Local grid operator, owns hydropower plants
Alvesta Energi	Municipal utility. Local grid operator, owns biomass CHP plants

the deregulation of the electric power market and has had different names: Höglands Energi, FyrstadKraft, Östkraft and, since 2009, Bixia AB.[11]

The company was established to create an alternative on the power market, with the local market conditions and demands as a starting point, and create new development paths. The company has grown but never lost its local character (Bixia, 2013).

According to the company Website:

> For us the future is renewable and locally produced—which separates us from many other power distributors. In the future the electric power doesn't have to come from large coal-fired power plants in Europe or from giant windpower plants. When we at Bixia mobilize power, it may as well come from a local farmer who utilizes his small power plant with support from our environment fund.
>
> We also believe in the power that is generated when many small actors cooperate toward a common goal. Therefore we keep our local offices in different parts of the country. We collect local power. Through us many small electric power companies can continue their activities in their local contexts, and give good advice so that our customers can make smart and enlightened choices.
>
> Today Bixia is the company that buys the largest share of locally produced electricity and it will increase even more in the future. (...) There are many ideas, but one thing is certain—to create the sustainable society we all opt for it takes a branch that dares to think along new lines.
>
> (Bixia, 2013, my translation)

The company started an affiliated windpower limited company, Bixia ProWin AB, in 2009 (see Section 6.8.4).

It offers its customers and the public to buy shares in this company, to contribute to a better climate:

> Our offer is a unique and new opportunity for the public to contribute to real change and at the same time get returns on their investment.
>
> (Bixia, 2013, my translation)

The interest in buying shares in windpower has been big, and the two issues made so far have sold out rapidly. The shares are offered only to company customers, not private consumers.

The motive of Bixia to use windpower is to create new business opportunities. This will strengthen the company's role on the electric power market and also its profitability Torbjörn Tengstrand, from Bixia ProWin AB, explains:

> The directives from the board state clearly that we shall focus on windpower. Normal demands on return on investments are 10 percent to 12 percent,

> but for windpower development, 6 percent to 8 percent is enough. To develop windpower is part of the company's business strategy. We get new customers, and all power is managed by our power trading company.
>
> (Tengstrand, 2013)

7.6.13 Motives of municipal actors

This municipal windpower investment boom has several explanations. The most important motive was the political target to create a renewable energy system. The strongest motive was that it was good business. It actually was better business for a municipal company to invest in windpower than for other actors on the market.

According to the electricity law, a power consumer that operates its own power plant and neither buys nor sells any power does not have to pay the electricity tax. This applies even if the power is transmitted by the power grid. So a municipal company that owns a wind turbine pays for the grid transmission and the self-cost price for the wind-generated electricity but no electricity tax (which corresponds to more than 25 per cent of the consumer price for electricity; the share depends on the current market price).

Since nothing is sold or bought, there is no VAT either (except for the grid transmission cost). On top of that, the owner receives certificates for the power produced. And on top of *that*, municipalities have their own investment bank, KommunInvest, which offers very good terms and low interest rates. There are other models of municipal ownership and motives for investing in windpower.

There are different reasons for municipal utilities to engage in and support cooperative windpower associations. One is political: to participate in the transformation of the power system to sustainable energy sources and give citizens the option to invest in renewables.

There is, however, also a business idea. The utility will manage the administration of bills, which they have the software and staff to do, and also supply the balancing power to the members of the cooperative. These have to become customers of the utility and will remain customers during the windpower plant's lifetime. The members on their side do not have to manage any bills, so this is a win-win situation (Larsson, 2009).

7.7 Motives for investments in windpower

The different ownership models have been developed by project developers to make it possible to finance and sell the windpower plants they are planning and building. They have developed models that are attractive for investors. The question is what motivates different actors to make these investments in windpower.

The answer to this question is simple: all of these ownership models could be expected to give good return on investment for the share owners in limited companies, for members in windpower cooperatives, and for private and municipal companies. This was the case when the investments were made. In 2012, this was unfortunately no longer true, due to low prices for electricity and for certificates (Lindahl, 2012).

Many of the private owners are farmers. Farming is a business, and it is not surprising that investments in windpower are evaluated just as any other investment. This is probably true also for other private investors.

There are several different motives to start windpower cooperatives. The initiators are often individuals, with a personal interest in renewable energy technology and a political ambition to create such a system. By engaging others with the same ambitions to become members of windpower cooperatives, they are able to finance such projects. These motives are probably still strong for many people today who buy shares and become members of cooperatives but no longer the only or the strongest one.

The cooperative ownership model was chosen since this was a good way to finance windpower projects in the early 1990s, when banks did not give any loans to such investments. Later on, members joined cooperatives to make business investments: to get electric power at a favorable price. Many small and medium-size companies, both private and public, joined cooperatives for economic reasons, which indicates that the price for electric power from a windpower cooperative was competitive.

Municipal utilities have different reasons to engage in and support cooperative windpower associations. They are governed by politicians, and they may want to offer citizens the option to invest in renewables. There is, however, also a business idea. The utility can manage the administration of cooperatives whose members have to become customers of the utility.

There are windpower limited companies of different characters. There are small local companies, and the ones from the early 1990s were formed for the same reasons as the early windpower cooperatives: it was a good way to finance windpower projects. The motive of the investors/shareowners was to develop a renewable energy system. Today, there are also some public limited companies listed on the stock exchange. All limited companies have according to their statutes the aim to optimize the profits for their shareholders, but this does not necessarily mean that this also is the motive of the shareowners.

Companies in other lines of business have several different motives to invest in windpower plants. The main motive of these other companies is to secure stable long-term power prices by producing it with windpower plants that have fairly predictable costs. This is a long-term strategic interest that motivates large investments in windpower plants. Some have additional motives: to strengthen their trade mark (IKEA, Wallenstam) and utilize land areas for new purposes (SCA). Business companies have to make investments that should be profitable.

The most important motive for municipal utilities is the political ambition to create a renewable energy system, but it has also been good business. Windpower can be used to produce power for own consumption and does not have to pay electricity tax (this applies to private companies as well). Municipalities also use more common models of municipal ownership described above.

The motives of the transnational power companies are to defend and expand their market shares and to deliver a good surplus to their owners. Since their share of windpower ownership is declining, the question is rather to find their motives to *not* invest in windpower in Sweden.

Vattenfall still makes its largest investments in countries outside Sweden and more in fossil fuel than in renewables. Vattenfall has motives (the government mandate) but obviously lacks the motivation to develop windpower in Sweden.

Norwegian Statkraft makes large investments in windpower plants in Sweden. The motive for this is to get a foothold on the Swedish electric power market. Fortum's motive to invest in the windpower plant at Blaiken is motivated by business interests and evaluated as a profitable investment. This makes Vattenfall's low interest to invest in windpower even more remarkable.

Notes

1 Näsudden on Gotland is one of the windiest sites in Sweden. Around 2005, there were eighty-one wind turbines operating on this cape. During the last years, a generation shift has started. Most small turbines from the 1990s have been replaced by larger ones, so the number of turbines has been reduced and the production has increased. The owners of the old turbines have become part-owners (shareholders) in these new large-scale windpower plants.
2 Unfortunately, tax authorities in different regions of Sweden interpreted this law in different ways, so in some areas a business profit tax has to be paid.
3 Prices and so on from 2010.
4 The factories are, of course, connected to the power grid, but the turbines are connected to the factories' internal grid "behind the meter."
5 Power company owned by the Norwegian state.
6 More than 51,000 forest owners in southern Sweden are members of the economic association Södra. They own just over half of all privately owned forest in the area as well as a group of companies that are successful in both Swedish and international markets. This gives them a market for their raw materials from the forest and at the same time provides the foundation for profitable forestry.
7 This company was established in 1996 after the deregulation of the electric power market.
8 In Sweden there is praxis since the early 1990s to create funds for local benefits so that local inhabitants who do not have a stake in the projects from land lease or ownership will get some benefits from the windpower plants. There are no laws regulating this, but according to praxis 0.2–1 percent of the gross revenue is funded for local benefits (Hela Sverige Ska Leva, 2012).
9 A tax on the members' discount in relation to the market price of electricity was introduced, which made membership less profitable (Skatteverket, 2008).
10 In some areas of Sweden, the grid is operated by cooperatives. These were once established to electrify some rural areas, which was financed by local inhabitants, mainly farmers and local manufacturing companies.
11 Bixia is the name of the goddess of dawn in Chinese (i.e., start of a new era).

8 The relevance of windpower ownership

In this chapter, the results of the investigation are summarized and put in a wider context by discussing how the findings fit the theories presented in Chapters 2, 3, and 4.

Two research questions remain:

- What are the advantages/disadvantages with the ownership models for a fast transition to a renewable energy system or, more specifically, for a fast development of windpower in Sweden?
- How can laws, rules, and regulations be adapted to promote these ownership models, given that they have advantages?

These questions will be discussed at the end of this chapter.

8.1 Summary of findings

This investigation of windpower ownership has shown that windpower plants in Sweden are owned and operated by many different categories: private investors, cooperative associations, limited companies, and so on. Windpower is used for distributed power generation, but it has also a *distributed* ownership.

8.1.1 Ownership structure

The share of different owner categories has varied quite a lot during the years since windpower development in Sweden started in the early 1990s. The share of windpower limited companies has doubled. Power companies and utilities had a large share in the very beginning, but the share has decreased to around 20 percent. The share of private and cooperative ownership is declining. Other companies have had a share just below 20 percent during this period from 1990 to 2012.

The share of power companies from the dominant energy system (DES) is small and declining. This could seem to indicate that a significant change of the ownership structure in the electric power sector is taking place. This is far from the truth. Even if windpower is developing rapidly in Sweden, it still has a rather modest share of the electric power sector, with less than 5 percent of the power production. This share will, however, increase.

The share of power companies and utilities of the windpower capacity in Sweden in 2012 was 24 percent. If power companies and utilities are separated, it turns out that the share of power companies has declined from 22 percent in 2007 to 11 percent in 2012 but that the share of municipal utilities during the same period has increased from 6 percent to 13 percent.

The share of the state-owned power company Vattenfall, which owns one-third of the power companies' windpower capacity (only 7 percent of the total capacity), is declining. This is surprising as Vattenfall, according to the directives from the government (implemented in 2009), should be one of the companies that *lead the development of an environmentally sustainable energy production.* In 2012, Vattenfall invested twice as much in power plants using fossil fuels (coal and gas) as in windpower, and all of the windpower investments were made in foreign countries.

In 2012, the share of non-utility generators was 76 percent; most of the windpower plants in Sweden are owned, operated, and managed by private companies that do not belong to the DES. Many different business models are used. These models have been continuously developed and adapted for different kinds of investors, to make it possible to finance windpower projects. Project developers have been very innovative and creative in this development of new ownership models.

New owners in the consumer windpower category are large industrial companies, and these will likely own an increasing share of windpower in coming years. Several large windpower plants owned by such companies are in the building stage, and even more have received all necessary permission. This will increase the share of "consumer power" very much in the next few years.

More than half of the windpower plants in Sweden are owned by windpower limited companies, many of which are small and local. The share of "community power" (i.e., power plants with local/regional ownership) is estimated to be at least 40 percent and the share of Consumer Power—around 6 percent.

8.1.2 Owner motives

The motives for investing in windpower differ. In the early 1990s, when windpower development in Sweden started, an investment in windpower was a "political statement," a demonstration of the will and possibility to create a sustainable power system. This was the motive of members of cooperatives, some local windpower limited companies, but also of municipal utilities that developed windpower plants.

This is still a common motive and is used by several companies for marketing windpower. Bixia Energi offers investors "to contribute to real change" and Telge Energi "to contribute to the development of windpower in Sweden and make coal-fired power plants disappear from the Nordic market," to mention two examples.

There are windpower limited companies of different characters. Most of them are small local companies, and those from the early 1990s were formed for the same reason as the early windpower cooperatives: to finance windpower projects.[1] The motive of the investors was to develop a renewable energy system. All limited companies have, according to their statutes, the aim to optimize the profits for their shareholders, but this does not necessarily mean that this also is the motive of the shareowners.

Companies in other lines of business have several different motives to invest in windpower plants. The main motive for the larger of these "other companies" is to hedge power prices (i.e., to reduce the insecurity of unpredictable and volatile power prices on the Nord Pool market). This is a long-term strategic interest that motivates large investments in windpower plants.

Some of these other companies have additional motives: to strengthen their trade mark (IKEA, Wallenstam) or utilize land areas for new purposes (SCA). Business companies have to make investments that should be profitable.

All of these ownership models could, when the investments were made, be expected to give good return on investment for the share owners in limited companies, members in windpower cooperatives and private and municipal companies. Many small and medium-size companies, both private and public, joined cooperatives for economic reasons, which indicates that the price for electric power from a windpower cooperative was competitive. This motive to make a good business investment, to realize good return on capital, or to get electric power at a favorable price has been a motive for all ownership categories.

Several municipal utilities have initiated cooperative windpower associations to give citizens an option to invest in renewables. There is, however, also a business idea. The utility can manage the administration of cooperatives, whose members have to become customers of the utility. After deregulation, this was a clever move to keep and maybe increase the number of customers.

The main task of transnational power companies, public and private, is to make high profits for their owners, so the motive is to make good business investments. The Swedish state-owned power company has also to implement the government mandate to *lead the development of an environmentally sustainable energy production*. However, Vattenfall still makes its largest investments in countries outside Sweden and more in fossil fuel than in renewables. Since Vattenfall's share of windpower ownership is declining, the question is rather to find their motives to *not* invest in windpower in Sweden.

Other state-owned power companies have shown more interest in developing windpower in Sweden. Norwegian Statkraft will make large investments. The motive for this is to get a foothold on the Swedish electric power market. The Finnish company Fortum's decision to invest in the windpower plant at Blaiken is motivated by business interests and evaluated as a profitable investment. This makes Vattenfall's low interest in investing in windpower even more remarkable. The perceived or real lack of fairness[2] of the electric power market is also a motive to invest in or to market windpower, has been mentioned by Telge Energi and O2 El.

8.1.3 Local-level energy policy

The share of windpower owned by municipal utilities and municipal companies has been increasing. This indicates that local politicians take more responsibility for the transition to a renewable energy system than does the government, which is responsible for the policy of Vattenfall.

Besides the pioneering municipalities of Falkenberg and Varberg, Dala Vind AB, owned and operated by a large number of municipal companies in the county of Dalecalia, stands out as an excellent example of how municipalities can realize a transition to a renewable energy system in Sweden in an efficient and democratic way.

These municipal utilities and companies have also promoted windpower cooperatives and given small and medium-size companies in their regions opportunities to become windpower owners. The new ownership model developed by Bixia AB is a good example of this.

Several informants stress the importance of municipal utilities, which are governed not only to optimize profits but to implement political ambitions. If there is a political will, municipalities can develop windpower and other renewables to create a local or regional renewable energy system.

8.1.4 Development of ownership models

Several informants mention that they started as members of windpower cooperatives. As members, they gained trust in the technology, learned how the economics worked, and gained knowledge to make new investments in windpower plants, to establish windpower limited companies or project-developing companies. The windpower cooperatives have had a role as plant schools for windpower developers and investors.

During most of the 1990s, it was not possible to finance windpower projects with bank loans. This had an impact on the ownership models that could be used. A windpower cooperative pays the investment up front, financed by the members, and needs no bank loans. Private investors and farming companies can get loans using other assets (farms, real estate, etc.) than the wind turbine as security.

When wind turbines, windpower plants, and investment costs increased, different models for mixed ownership were developed. The use of such "umbrella companies" made it possible for these different ownership categories to continue to invest in windpower plants.

Many windpower plants are also public–private partnerships. Municipal utilities have played a decisive role in the development of such partnerships by initiating windpower cooperatives and by offering local small and medium-size companies to invest in projects developed by them. This has been a very successful concept, initiated by Falkenberg and Varberg and further developed by Dala Vind AB and Bixia, where several municipal utilities cooperate and develop windpower plants that are financed by local public and/or private investors. Several project developers, such as Eolus Vind AB, O2, and others, also try to involve municipalities and public companies in their projects.

A rather new model for cooperation has been joint ventures. These are used for large-scale windpower development. Blaiken Vind AB, a joint venture formed by the municipal utility Skellefteå Kraft AB and the Finnish power company Fortum, and Statkraft SCA Vind AB, formed by Norwegian Statkraft and the forestry company SCA, are two examples of such joint ventures, which will become owners of a large share of windpower in coming years if their development plans are realized.

Another interesting development is that several windpower owners and operators have formed their own power-trading companies. Telge Energi is a power-trading company that has windpower electricity as a product. It now offers its customers to become members in windpower cooperatives that they have initiated. Bixia has a power-trading company; Dala Vind AB is affiliated to the power-trading company Dala Kraft AB; the company O2 has its own power-trading company; and so has the real estate company Wallenstam.

8.2 Networks with windpower

How do these findings fit the theory on large technical systems and others that were presented and discussed in Chapters 2 through 4? These theories may help to explain why the state-owned power company Vattenfall's share of windpower ownership is declining, and the company's motives to *not* invest in windpower in Sweden.

8.2.1 Large technical systems

Hughes found in his book *Networks of Power* that the electric power system had become mature; no more radical innovations could be expected, and the electric power system had become integrated in national transmission grids. This happened in the early 1930s.

Some technical components have been improved: the capacity of power plants as well as transmission and distribution lines have increased; and new types of power plants, such as nuclear reactors, wind turbines, and PV-cells, have been introduced. The electrical power system has been improved but, from a technical point of view, it is basically the same.

Hughes divides the development of the electric power system into four phases—*invention and development*, *technology transfer*, *system growth*, and *momentum*—that can overlap. Högselius and Kaijser add the phases *saturation* and *stagnation* to this model.

In Sweden, the electric power system has reached saturation, as in some other European countries as well, but in most other countries of the world, especially in the developing countries, the systems are still growing very fast.

The development and the integration of windpower in the electric power system have passed similar steps. It started with *invention and development*, where inventor-entrepreneurs played a dominant role. Then *technology transfer* from one region and society to another took place. Windpower technology was exported from Denmark to California in the United States, to India, Germany, and Spain, and finally to most countries in the world. Entrepreneurs, organizers of enterprises, and financiers played a dominant role.

A large technical system such as the electric power system consists of three interacting parts: technical, economic, and political. The technical systems are almost identical, but there are large differences between countries in the economic and political parts of the electric power systems. When windpower developments in different countries are compared, it is also obvious that variations in geographical, cultural, managerial, and entrepreneurial character (i.e., the socio-political context) play important roles.

In China, with most windpower capacity in the world, the development is governed by the central government, which sets targets and allocates means to implement them. In Spain, large power companies from the DES develop, finance, and operate windpower plants, while in Sweden this role is performed by small private companies and some municipal utilities, as the state-owned power company Vattenfall prefers to make investments in large offshore windpower plants in other countries in Europe.

Reverse salients[3] have a different character than when the electric power system was developing. There are two main reverse salients in the electric power system today. The first is nuclear power, which after the Fukushima catastrophe again has been proved to have a very high risk.The second is the use of fossil fuels that cause global climate change. A third reverse salient, which now has been technically solved, has been the adaptation of wind turbines to enable a large-scale integration of windpower in the electric power system.

8.2.2 *Integration of large technical systems*

The integration of windpower in the electrical power system has become possible owing to the fast development of other technical systems: power electronics, and information technology. Components from those systems are integrated in the electric power system and will also make it possible to develop a so-called smart grid (a *smart power system* is a better term). This technology has also been a precondition for the deregulation of the power market, as it has made it possible to measure inputs and outputs from the system in detail, transmit this information quickly, and thus coordinate many different and independent actors in the system.

The most radical change in the electric power system since the 1930s has taken place in the political and economic subsystems by the deregulation of the electric power market. This has changed the technical as well as the economic management of the system, the business models, and the political governance. It has caused the most radical change of its style; the actors' aims are to *optimize the company profits rather than to minimize the cost of electric power for the consumers.*

Other very interesting developments on the large technical system level are the relations between the electric power systems and the heating systems and transport systems, respectively. Heat, and fuels to produce heat, have been part of the electric power system from the start. To use electricity to produce heat is a more recent development. Direct electric heating was introduced on a large scale in Sweden, to create a market for the surplus production from nuclear power plants. At the same time, oil was replaced by a cheaper and cleaner source of energy. During the last decades, the use of more efficient heat pumps has expanded. There are also many CHP plants operating, which produce both heat and power. Systems with energy storage for heat (and cooling) are also increasing in number.

Electricity has also been used in the transport system from the start, for urban trams. Traction was as important as lightning for power companies at the turn of the last century. Later, electric trains replaced steam locomotives on the national railways. Now electric cars have started to appear on the market. If this market expands, private transport will also be integrated. The electric power system will expand, and to make this in a way that will be sustainable, the electric power has to come from renewables, which will increase the demand for windpower.

8.2.3 Vattenfall and windpower development in Sweden

Vattenfall actually owns and operates several very large windpower plants; most of them are offshore windpower plants in Denmark, the United Kingdom and other countries. In Sweden, Vattenfall has one offshore windpower plant (Lillgrund), one large onshore plant in northern Sweden, Stor-Rotliden, and a few more. Vattenfall actually invests in Sweden, but its share of windpower capacity is decreasing.

According to Hughes, the management structure of a power company depends on the character of the artifacts working in the system, and the other way around, and management often chooses technical components that support the structure of management.

This could explain the resistance to windpower development from power companies that are parts of the DES, up to about 2006.[4] Power companies are used to centralize systems with large power plants, and the company structure is adapted to manage such systems, not systems with distributed generation.

In 2006, wind turbines had become larger and, by developing windpower plants with many large wind turbines at the same site, onshore and offshore, windpower could be integrated in such centralized power systems.

It is, of course, not more economical to own and operate offshore windpower plants in Britain than windpower plants on land in Sweden. The cost price for offshore is more than double (see Figure 2.6).

The reason that Vattenfall has chosen to invest in windpower in countries other than Sweden is probably because the support systems give a higher and safer return on investments.[5] Another reason could be that it is easier and faster to build such large projects in countries other than Sweden, where the permission process still is a very time-consuming high-risk venture.[6]

Path-dependence and lock-in are plausible explanations of the lack of interest and investment up to 2006. Such lock-in is caused by several mechanisms: cognitive routines that blind engineers[7] to developments outside their focus, sunken investments in machines, infrastructure, and skills, vested interests, and so on (see Section 2.2.1 for the complete list).

During Vattenfall's expansion on the European market, the company invested in coal-fired power plants, coal mines, and old nuclear power plants. In the summer of 2013, Vattenfall had to make a depreciation of its assets with 30 billion SEK.

Several transnational power companies had encountered the same economic problems owing to the economic depression in Europe and competition from renewable energy sources, which have developed rapidly in Germany, Denmark, and other countries. The decision by the German government to close down old nuclear power plants, two of them owned and operated by Vattenfall, has also caused great losses for the company (Forsström, 2013).

This is a good example of lock-in; to make such huge investments in old technology is a consequence of blindness to developments outside the focus of a company in the DES.

The development of a renewable energy system has to be initiated by actors outside of the power industry, according to Henrik Lund. The development of windpower in Denmark, Germany, and Sweden confirms this.[8] Lund also stresses that different systems have to be analyzed not only from a strict technical and business economy point of view. The institutional frameworks, like taxes, tariffs, and the like, as well as political goals, such as high employment and balance of payment, should be included.

One difference between Germany, where windpower development has been very fast, and Sweden, with a much slower development, is that Sweden has a large state-owned power company, Vattenfall. This company has a great influence on energy policy in Sweden; it is a part of the state. Vattenfall has large sunken costs in power plants and vested interest in the DES. Alternatives, which are better from an environmental and also socioeconomic point of view than those proposed by the power industry, have not been able to get on the political agenda.

The government has also delegated to the power industry power to set the rules of the power market and promotion systems, and these are consequently constructed to fit the DES. It is well proved that fixed-price systems such as feed-in tariffs are preferred by investors (see Section 2.2.7). Still, a quota-certificate system was introduced in Sweden.

In Denmark and Germany, promotion systems have been designed by politicians, in close dialogue with the national windpower associations, and have been opposed by the power industry.

8.2.4 *Windpower as part of the dominant energy system*

In Section 2.2.11 (Transition and ownership), the following question was posed: What will happen when windpower has become a mainstream technology, which has happened during the last few years? Will the actors from the DES take over the ownership of windpower plants and become a dominant player, or will these new independent owner categories continue to play a significant role?

So far, actors from the DES have not taken over. Large power companies have so far mainly invested in large offshore windpower plants, which still have a marginal role in the power systems. The trend in Sweden moves in the opposite direction: the power companies' share of windpower ownership is declining.

Municipal utilities, which also can be considered as part of the DES, have, however, played an important role and promoted the development of windpower in many different ways, using innovative ownership models to enable citizens and local enterprises to participate with their resources to create a renewable energy system.

8.3 Advantages and disadvantages

Now it is time to address research question 4: What are the advantages and disadvantages with the ownership models for a fast transition to a renewable energy system or, more specifically, for a fast development of windpower in Sweden?

In Chapter 5, it was shown that many different ownership models have been used and that these have developed into new refined and innovative models during the period when windpower has been developed in Sweden. The driver for this development has been the project developers' efforts to finance windpower projects. Models have been adapted for different categories of investors: power consumers, farmers, small and medium-size enterprises, and so on.

As power companies from the DES have not been very interested in investing in windpower, these ownership models that make it possible and simple for other kinds of investors to finance windpower development have obviously been an advantage for a fast transition to a renewable energy system in Sweden.

There seems to be a direct relation between ownership type and degree of local support for windpower developments (see Section 4.2.9). Local support of wind turbines will depend on the extent to which some of the economic benefits of the development are given to the people who have to live with the impacts of these turbines (Sperling, 2007).

Environmental, economic, and social commitment may all be reasons for people to invest in windpower. Community projects make it possible to combine these motives. There are many reasons for people in local communities in different parts of the world to support windpower and many reason for them to oppose windpower projects. Ownership seems to be an important aspect in this regard. Finally, issues of fairness and equity are important factors for support of windpower projects.

Several of the ownerships models used in Sweden, those that can be classified as community power projects, will thus increase local support for windpower as well as local investments and benefits.

These ownership models have obvious advantages by making it possible and easy for different kinds of small investors to finance a transition to a renewable energy system. They have been a *necessary condition* to start this transition to a renewable energy system. This advantage is even more important as the large power companies in the DES so far have made only symbolic investments, and their share of windpower ownership is declining. Any apparent disadvantages have not been identified.

8.4 Adaptation of laws, rules, and regulations

The final research question was: How can laws, rules, and regulations be adapted to promote these ownership models, given that they have advantages?

Non-utility generators have played and will likely continue to play a crucial role for the transition to a renewable energy system in Sweden. Considering this, rules and regulations need to be adapted to encourage and promote this development. The most crucial factor is the promotion system.

Before more specific suggestions concerning laws, rules, and regulations for ownership models are made, it is important to point at the problems with the promotion system for renewables used in Sweden. To have a well-working and fair promotion system is a *necessary condition* for a continued transition to a renewable energy system for most of these owner categories.

8.4.1 The certificate system

Investor security is a reasonable demand for farmers, companies, and consumers who invest their money to contribute to the transition to a renewable energy system, which the Swedish government and parliament have agreed to develop. In 2012 and 2013, the power prices and certificate prices were so low that windpower owners could not make ends meet.

The annual economic returns from windpower depend on three factors: the wind resources, the market price for electric power, and the price for certificates. These three variables are volatile. If all three are low during the same time period, revenues will be very low. This was the case in 2012 and 2013 for windpower owners, who made their investments a few years earlier (2008–2010). The owner of a 2-MW wind turbine suffered a loss of 2 million SEK (Lindahl, 2012).

Several project developers also encountered severe problems and had to cut down staff, and the value of shares on the stock exchange for windpower companies sank radically.[9] The entire windpower business was in a serious crisis in 2012–2013, as revenues did not cover the costs. Those who made investments in windpower in 2008–2010, when prices were high (power, certificate, and wind turbine prices), have lost at least 25 percent of their money. This is not a good credit for a promotion system. The interest for new investments in windpower in 2013 was very low.

Investors and financial institutes want to minimize risks. The preferred system is one that sets fixed prices, at least for the period it takes to pay back the investment. Such support systems, with feed-in tariffs, are the most commonly used and have proven most successful (see Section 2.2.7).

In Sweden, the government asked a consultant, former vice CEO of Vattenfall, to develop a new promotion system for renewables in Sweden. This was done with great haste, without consulting different actors in the power system, which otherwise is common practice in Sweden (Högselius & Kaijser, 2007).

The certificate system that was introduced in 2003 is a system of financial engineering, a quota system with a virtual currency—certificates—wherein the value is set by supply and demand on a certificate market. A similar quota system has been used earlier in the agricultural sector for milk production. The purpose of that quota system was to *limit* the production of milk.

The certificate system was developed by a representative of the DES, which is a conflict of interest. However, with a state-owned power company, the government itself becomes a part of the DES. The government and Vattenfall have vested interests in the current power system and the nuclear power plants. If windpower and other renewables become too competitive, this will make it possible to close down the old nuclear power plants, which is not in the interest of the companies of the DES. Windpower also lowers the market price for electric power (see Section 3.4.7). From the DES power companies' point of view, the certificate system is perfect.

Wind turbines are cheaper and thus more cost-efficient in 2013 than they were in 2009. This is, however, mainly owing to changes in the exchange rate.[10] The more windpower that is installed, the lower will the price of certificates become. The problem is that it works retroactively and punishes early investors. When performance increases, the "rewards" are reduced.

A promotion system that punishes those who invest in power plants that the promotion system is supposed to support may feel cheated—by the government and the state.

No one in the windpower business is in favor of replacing the certificate system with some other promotion system, as this will cause a standstill of development until a new system is implemented and tested. The easy way out is to modify the system.

The price of the certificates is not decided by the "market," as some proponents claim (see Section 2.2.7), but by the size of the quotas, which is a political decision. In 2012, the Swedish Windpower association demanded in an open letter to the government that the quotas should be raised to get the windpower business out of this crisis (Lindahl, 2012). The government refused and postponed the revision of the quotas to 2015, according to the original plan.

Windpower investors make risk analyses, and the risk that power prices and certificate prices would sink to levels too low to make ends meet[11] is classified as a political risk. And this risk has been estimated to be close to zero. The presumption has been that politicians would prevent such a thing from happening. Now when it happens, it is obvious that this risk has been underestimated.

A simple way out of this "system error" would be to set a "floor" on the certificate prices (i.e., to guarantee a minimum price). This minimum could be adapted and reset each year but, for an investor, it has to be on the same level for the fifteen years they get certificates. This will make investments safer. During the first few years of the certificate system, such a floor was used, so it is not hard to re-implement. It is reasonable to also set a ceiling, to avoid windfall profits if certificate prices rise to a very high level.

Without such a reform, there is a high risk that investments by the ownership categories that develop windpower in Sweden will stop. The only actors that could afford to take such risks are the large power companies.

8.4.2 Unambiguous rules and regulations

Many rules and regulations concerning the ownership models used for windpower plants are ambiguous. The consequence of this is that the economic conditions for investors can be changed—after they have made their investments. This problem can be solved if rules and regulations concerning windpower ownership are made clear, by political decisions, instead of interpretations of, for example, tax regulations by authorities. A simple solution to avoid this insecurity would be to formulate a windpower law, just as has been done in Denmark.

Many rules and regulations, not least concerning taxes, are ambiguous. The project developers, who design business models, make their interpretations after consulting the local tax authority or other experts. After starting the windpower plant, some other actor may question the interpretation and send an inquiry to the central tax authority (Skatterättsnämnden), which makes an official interpretation of the law.

These laws were not specifically made for windpower ownership but for cooperative economic associations, enterprises, limited companies, and so on. Is a wind turbine a building or a machine? The answer will decide the depreciation time. Is an economic association a producer or a consumer cooperative? The answer will decide whether company tax should be paid or not. Such questions are difficult to interpret in an unambiguous way, and it takes time for the central tax authority to find the right answers.

Some reinterpretations have been made by the central tax authority. The one that has had most impact is the decision that members of consumer windpower cooperatives have to pay a tax on the difference between the price they pay for the power from the windpower cooperative and the market price for electric power. This is clearly stated in the law on economic associations: members who take a product from the association for their own use and get a rebate on the actual price of the product have to pay tax (according to specific rules) on the difference between the rebated price and the actual market price.

To apply this law paragraph on windpower cooperatives is, by most people, considered as unfair, as members get a lower price because they have made an *investment* and, on top of the "rebated" price, members have a capital cost, which makes the actual price higher.

Fairness and common sense are, however, not considered in legal matters. Swedish law practice is unique, as it makes judgments by a strict interpretation of the law: a semantic analysis. No common sense, no evaluations, only literal text interpretation (Töllberg, 2005).

In the law, there is an exemption for housing associations, which is a strict parallel. Members invest in an apartment, and rent equals capital cost of the association plus operation and maintenance. The investment to buy the apartment is a personal debt.

This interpretation of the law "killed" the market for windpower cooperatives, which now have difficulty in recruiting new members.[12] This could also be the reason that the share of windpower cooperatives has declined. The simple solution to this problem would be to just add an exemption for windpower cooperatives from this tax in the law on economic associations.

An even better solution would be to introduce a specific *windpower law*, as they have in Denmark. In such a law, all rules and regulations for these windpower business and ownership models could be specified by politicians instead of leaving it to the interpretations of the tax authorities.

With clear-cut rules and regulations and a reliable trustworthy promotion system, these windpower owners can continue to play a leading role for the transition to a renewable energy system in Sweden.

Notes

1 A common model was that, for a share of 250 SEK, the shareowner was committed to also lend 1,000 SEK to the company, which then was paid back with interest according to a specified plan.
2 Elinor Ostrom identifies fairness and trust as important variables to create robust systems for management of common pool resources (Ostrom, 2005). These variables are probably important also for the management of the electric power market.
3 Components that cause friction in the system.
4 The year when resistance turned to acceptance has varied for different companies and countries.
5 Fixed and high power price for many years, according to the accepted tender.
6 This will be the topic of my next book.
7 Economists and managers should be added.
8 Spain and China are interesting exceptions.
9 The value of shares on the stock market for Eolus Vind AB decreased by more than 40 percent and for Arise Windpower AB by more than 50 percent from 2010 to 2013.
10 The price for one euro has been reduced from around 11 SEK in 2008/09 to 8.50 SEK in 2013.
11 When revenues from sale of power and certificates are lower than capital costs plus costs for operation and maintenance.
12 This extra tax is actually rather low and does not make membership in windpower cooperatives unprofitable.

9 Future outlook

In this final chapter, conclusions are drawn, and some ideas for future research are presented.

9.1 Conclusions

The transition to a renewable energy system in Sweden by development of windpower plants has been conducted by non-utility generators and some municipal utilities. Investments have been made by private persons, farmers, small and medium-size enterprises, either directly or by shares in wind limited companies. Windpower cooperatives have played an important role as a plant school for project developers and local investors. Project developers have designed ownership and business models for these different owner categories to make it possible to finance windpower plants.

Recently, some large industrial companies have joined this development, often by forming joint ventures with transnational power companies. Also, some public and private companies in other lines of business, such as real estate companies, have invested in windpower plants to secure stable long-term prices for their electric power consumption.

Several municipal utilities have played an important role in the development of windpower. These have had some political ambitions, have been used as instruments to implement a local and regional energy policy, and have taken responsibility for the transition to renewable energy system that has been set as a target by the Swedish parliament. These municipal utilities have also initiated some public–private partnerships and promoted and enabled their citizens and local enterprises to invest in windpower.

The state-owned power company Vattenfall has played a surprisingly limited role, although it has been given the task to lead the development of an environmentally sustainable energy production in Sweden.

The conclusion is, thus, that private investors and municipal utilities are the main actors who implement a transition to a renewable energy system in Sweden. As a consequence of this, the role of these actors should be acknowledged by the government, and the necessary conditions for these ownership categories to continue to play an important role should be secured and improved.

Since 2012, the revenues from sale of electric power and certificates from windpower plants have been very low, too low to make ends meet for those who made their investments a few years earlier. The owner categories that have driven the development of windpower in Sweden are no longer able to finance windpower projects with such high economic risks.

Windpower investors have underestimated the political risk of the current government's political ideology to let the "market" alone set the prices for electric power and certificates. This is, however, an illusion: the electric power market is an oligopoly market (see Section 3.4.9), and the certificate prices are regulated by political decisions about the size of the quotas.

The problem with economic risk for non-utility investors can be solved by setting a guaranteed minimum price for certificates. During the first few years of the certificate system, there was such a minimum price to reduce risk for investors, so it could easily be reintroduced and combined with a maximum price to prevent windfall profits when certificate prices are very high. Otherwise, investments by the ownership categories that develop windpower in Sweden will stop.

Another problem is the ambiguity of rules and regulations, which also is an investment risk. These rules should be made clear by politicians, for example, by implementing a windpower law such as they have in Denmark. Conditions should, of course, be favorable for those who invest in windpower, and such favorable conditions could be justified as a part of the promotion system.

9.2 Future research

A very obvious and interesting task would be to follow up the shares of these ownership categories in coming years. The current economic crisis for the windpower business may result in significant changes, and some large windpower plants developed in 2013 could do so as well. It would also be possible to increase the resolution of the ownership statistics, by specifying all part-owners in "umbrella companies."

A similar investigation of ownership and business models used for other renewable energy sources, such as PV-cells, biogas plants, and so on, would also be interesting. PV-cells installed on private and public buildings and biogas plants on farms could have a large impact on the ownership structure of the electric power system.

To compare ownership and business models used in different countries and to investigate which models could be exported or imported could also be worthwhile and promote the transition to a renewable energy system. The consumer cooperative model is, for example, used only in Sweden but could be used in other countries as well.

To study the role and impact on the power market of independent power traders affiliated to owners/operators of windpower plants and other renewables would also be very interesting (see Section 8.1.4).

In Section 2.1.8 (Windpower as a social-ecological system) the framework developed by Elaine Ostrom and her colleagues for analyzing social-ecological systems is described. It would be interesting to investigate whether this framework can be used to analyze the electric power system and its relation to the global energy system and what patterns of interactions and outcomes are likely to result from a particular set of rules for the governance, ownership, and use of windpower in a specific technological, socioeconomic, and political environment.

This would be a challenge and, if the answers could be found, it would be a very useful guideline for policymakers who develop policies for the transition to a renewable energy system.

Appendix A

Interviews

Interviews were made during different periods and contexts. To get material for the manuals for farmers, windpower cooperatives, and municipalities, interviews were made with different actors by me or an assistant. These were semi-structured interviews, starting with "Tell me about windpower ..." and with some control questions to get information on facts and figures and the motives for investing in windpower.

I have reused some of this material by quoting these books. I made some additional interviews in 2013, using the same principle, with some additional control questions on ownership models. All interviews were recorded and then written on paper. These interviews are quoted in Chapter 5, *Ownership Models in Sweden* (2), and Chapter 7, *Windpower Owners' Motives* (19). What follows is a list in order of appearance.

Informant	*Interview by*	*Year*	*Note*
Andreas Wickman	Tore Wizelius	2013	By telephone
Thomas Sirland	Tore Wizelius	2013	By telephone
Åke Ahlsten	Gunilla Britse*	2006	
Per Ericsson	Tore Wizelius	2009	
Christer Larsson	Carin Persson**	2010	
Håkan Svensson	Carin Persson	2010	
Linda Magnusson	Tore Wizelius	2010	
Berndt Bolms	Tore Wizelius	2009	
Åke Westberg	Tore Wizelius	2013	By telephone
Bo Strandberg	Tore Wizelius	2013	By telephone
Johan Risholm	Carin Persson	2010	
Jan Åke Jacobsson	Carin Persson	2010	
Jan Högberg	Carin Persson	2010	
Inge Östlund	Carin Persson	2010	

Staffan Larsson	Tore Wizelius	2009	
Erik Persson	Tore Wizelius	2009	
Lars C Johansson	Tore Wizelius	2009	
Ulf Eriksson	Tore Wizelius	2009	
Åke Pettersson Frykberg	Carin Persson	2010	
Stefan Sundelius	Tore Wizelius	2013	By telephone
Torbjörn Tengtrand	Tore Wizelius	2013	By telephone

* Gunilla Britse, Gotland University, co-author of the first farmers' manual published in 2006.

** Carin Persson, engaged as an assistant to conduct interviews for the manuals for windpower cooperatives and municipalities.

Appendix B

The electric power system

The electric power subsystems

The electrical power system can be divided into three interrelated subsystems:

- technical system;
- legal/political system; and
- economic system.

First, these systems are briefly described. The relations between these levels are discussed afterward. It is the Nordic electric power system that is described. The technical system works the same way in all electrical power systems over the world, as it is governed by the laws of nature, and common standards are applied, although grid frequency, voltage levels, and other technical parameters may differ. The legal/political system and the economic systems, however, differ in different countries. The economic system and the Nord Pool market are described in Appendix C.

These three subsystems interact with one another, as illustrated in Figure 2.1 in Section 2.1.6.

The *technical* system consists of three different components:

- power plants that produce the electric power;
- power grid, which transmits power to consumers; and
- electrical equipment used by power consumers.

Electricity is not an energy source but is utilized to transport energy in the form of electric power from power plants to power consumers. The electric power grid constitutes a practical and cost-efficient means to transport power from plants to consumers, who with different types of equipment transform the electric power into light, heat, or mechanical work.

In Sweden and most other countries, the backbone in the power system consists of a *national transmission grid*, with a voltage level of 400 or 230 kV that is managed by a transmission system operator (TSO). In Sweden, this role is filled by the state-owned public utility Svenska Kraftnät (Swedish

National Grid). The very-high-voltage transmission grid is used to transmit large amounts of power over long distances.

The next level is the *regional grid*, with a voltage level of 130 or 70 kV (in Sweden) that transmits power from the national transmission grid to the local *distribution grids*. These have lower voltage levels—40, 20, or 10 kV—and are utilized to distribute power to factories, households, and other consumers. Before the power enters the consumers' low-voltage grid, the power is transformed to 690V for industry or 400V (230V per phase) for other electric power consumers.

To transmit large amounts of power over long distances, it is most cost-efficient to use as high voltage as possible. With all transmission of electric power, losses (in the form of heat) are unavoidable. How large these losses will be depends on the dimensions, lengths, and distance of the power lines. In the Swedish power system, around 10 percent of the power that is fed into the grid will be lost. The TSO is actually the largest power customer in Sweden, since the power that is lost has to be bought and paid for.

In Sweden, the power system is based on a few very large power plants: hydropower plants in the north of Sweden and nuclear reactors in the south. There are also some CHP plants that produce heat and electric power simultaneously and some gas turbines that are used for peak power when the loads are extremely high during cold winter days or to balance sudden increases in power consumption.

Electric power consumption

The electric power consumption in Sweden is in the range 140 to 150 TWh/year (Energimyndigheten, 2010). It has been fairly constant at this level since 1987 (Högselius & Kaijser, 2007). During different periods of the year, power is also imported or exported.

The power consumption is always changing, during each day and also during the year. The need for power is, of course, less during the night, when factories are idle and people are sleeping. In the morning, when people prepare breakfast and factories, shops, and offices start to operate, power consumption increases fast, and in the evening it decreases again, interrupted by an increase at dinner-time (see Figure B.1).

In Sweden, where many dwellings use electric heating, electric power consumption is much higher in the winter than in the summer. The peak usually occurs in January or February, when it is cold all over the country and the wind is blowing as well. Then the power demand is at its highest; all power plants run at full capacity, and some power will also be imported (see Figure B.2).

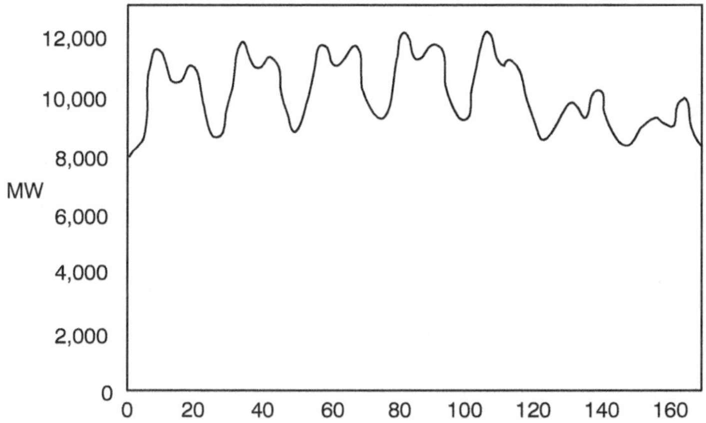

Figure B.1 Variations of power consumption: diurnal and weekly. The diagram shows the variations of load during a normal week. The vertical axis shows the power demand (consumption) and the horizontal axis the hour of the week, starting 00.00 Monday. The peaks are significantly lower on weekends. Illustration (Söder, 1997)

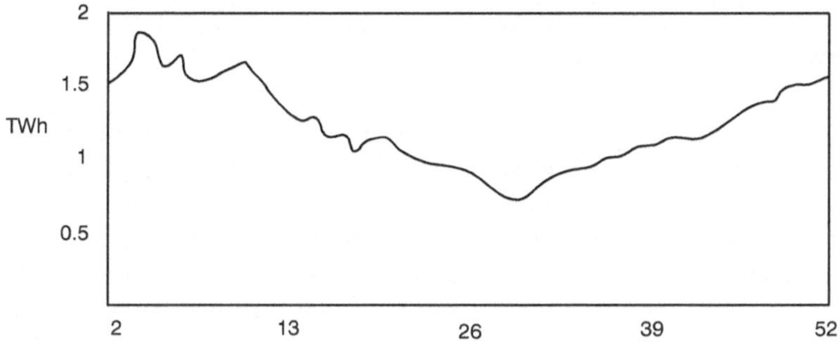

Figure B.2 Annual variations of power consumption. The diagram shows the weekly variations of power consumption in a region in Sweden during one year. In a cold week in winter, power consumption is more than twice as high as during the summer holiday weeks Illustration (Blomqvist, 1997)

Balancing power production and load

Electric power is a perishable commodity; power production in the grid has to match the power consumption[1] at all times. If not, the frequency will change. If the power production is larger than the consumption, the frequency will increase; if consumption is larger than production, the frequency will decrease. In an electric power system, frequency is universal (i.e., exactly the same at every place in the system). The voltage level is local and can vary at different places (nodes) in the system. When there is a mismatch between supply and consumption (load), electric equipment can be damaged or the grid breaks down—a blackout occurs. However, this happens very seldom, and the grid operators know how to manage the power system and to handle the balance of supply and demand.

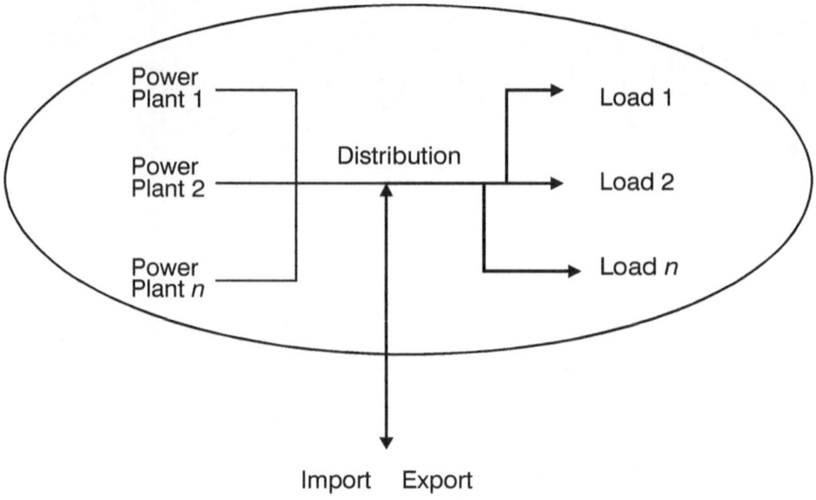

Figure B.3 The electric power system. The technical system has its limits at the borders of the countries in the Nordic power market. Within these borders there is an identical frequency—50 Hz—in the system. It is, however, no problem to import or export power to/from the system

Different power plants have different roles in this system: to supply *base load, power regulation*, or *peak load.* In Sweden, nuclear power plants are used for base load; they run day and night most of the year, and it is a slow and complicated matter to regulate the power output of a nuclear reactor. Hydropower also supplies some base load but is also very easy to regulate, and hydro power plants are used for this purpose. Gas turbines can be started within seconds and are used for peak power and when there is a need for very fast power regulation. Nowadays, peak loads, or rather power on the margin, are often handled by import of power from neighboring countries. It is also possible to regulate power consumption on the demand side, for example, by turning some large industrial loads down or off.

To be able to follow changes in the load (consumption), power reserves have to be available. *Primary reserves* are power plants with very short start-up times, from 1 second to 1 minute; these are used for fast regulation. The power plants used for this are gas turbines and hydropower. These are replaced by *secondary reserves*, power plants with longer regulation times (10 minutes to 1 hour) and with lower operational costs. When the secondary reserves increase their production, the primary reserves are replaced.

To prevent a blackout if a failure occurs at some large power plant so that it has to close down at short notice, there is always a *disturbance reserve* in the power system as well. This disturbance reserve is dimensioned after the largest power plant that could trip off. In the Nordic power system, the disturbance reserve is 1,200 MW, the size of the largest power plant (a nuclear reactor) in the power system.

Windpower in the electric power system

Wind turbines are quite small power plants that should be sited where the winds are strong. Therefore, they are often located at the periphery of the grid, where the grid is weak. One or a few turbines can always be connected directly to the distribution grid, but large windpower plants have to be connected to the higher voltage level on the regional grid.

The power output from wind turbines varies with the wind speed. Wind energy cannot be used for base loads, as the production varies. It cannot be used as regulating power, as the energy from wind turbines cannot be increased with demand. It cannot be used as peak power either, for the same reason. From the electric power system point of view, windpower works in a way completely different from conventional power plants. Wind turbines are usually connected to the distribution grid, the same power lines that consumers (households, industries) use. They function as distributed power plants.

The grid operator can handle the variations of power produced by wind turbines in the same way as variations in load (power consumption). It is simply treated as negative load. When the wind speed decreases, the power from other power plants, for example hydropower, will be increased. As long as the share of windpower is below 10 percent of the total power production, the intermittency is no problem (see Figure B.4).

Power companies and grid operators are not used to this (at least not in countries and regions where windpower development is in its infancy). However, there is no problem to integrate quite a lot of windpower in the power system.

When the windpower penetration (wind turbines' share of the power production) reaches a level of 10 percent, the grid and power system may need to be adapted to windpower. During the last five years, countries such as Denmark, Germany, and Spain have crossed this threshold. In

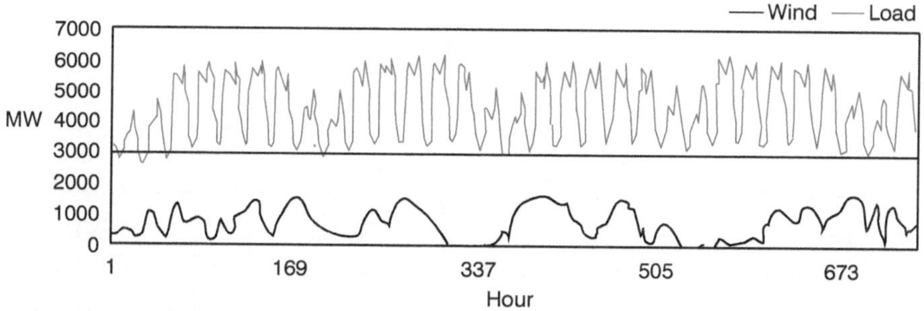

Figure B.4 Windpower variations. This diagram shows how loads (consumption) and windpower production varied in January 2000 in Denmark. The variations are handled just as ordinary load variations. When wind turbines produce much power, the contribution from other power plants is reduced, and vice versa (Holttinen, 2004)

some regions, windpower penetration is higher than that, like the German Länder Schleswig-Holstein and Nordfriesland (Gipe, 2011), but they are also integral parts of a much larger power system, which can absorb this production.

There is, however, no problem to use a much larger share of windpower in the power system. Power systems have good capacities to regulate power to keep supply and demand in balance; this is done every minute of the year, with or without windpower, and there are always power plants with reserve capacity in the power system.

Management of the technical system

The TSO has the overall responsibility for the operation of the electric power system. In Sweden, Svenska Kraftnät is the TSO and also operates the very-high-voltage transmission grid. The regional grids can be operated by different companies in different regions, such as Vattenfall, Fortum, and E.ON. The local grid operators, owned by large power companies or municipal utilities, manage the distribution grid.

The principle for the power system is that the production with low operating costs is run all the time and the production with higher operating costs are dispatched when demand increases, according to the so-called *merit order system*.

In a deregulated market, the task of balancing supply and demand is distributed to balance responsible players (BRP) and power trading companies. All production and consumption goes through a trader who has a contract with a BRP. These companies (BRPs) give production schedules to the system operator (TSO) one day ahead, but they can change the schedule up to the hour of delivery. During the operating hour, the responsibility is moved to the TSO.

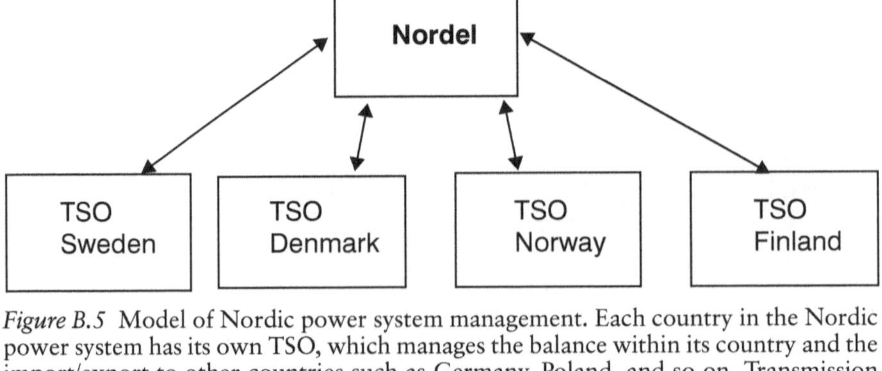

Figure B.5 Model of Nordic power system management. Each country in the Nordic power system has its own TSO, which manages the balance within its country and the import/export to other countries such as Germany, Poland, and so on. Transmission between countries is managed by the common organization Nordel

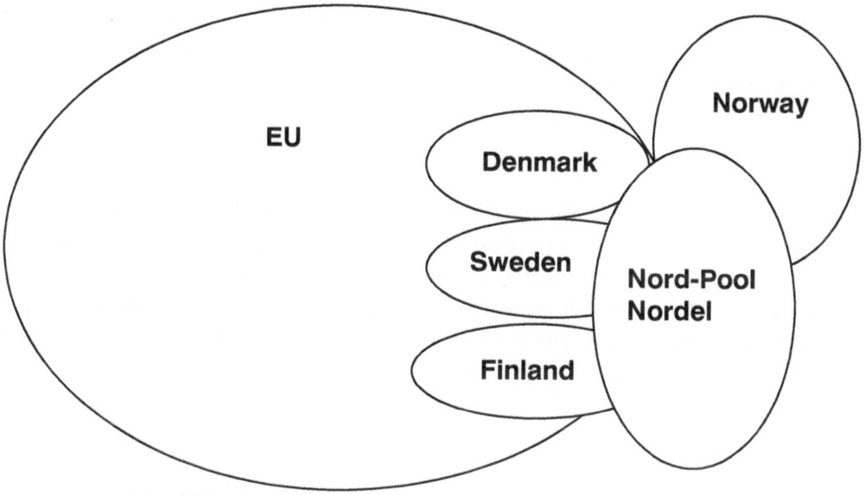

Figure B.6 Legal/political systems. There are international treaties, such as the Kyoto protocol, EU directives, and national laws, that set the framework for the energy policy. The rules for Nordel and Nord Pool are agreements between the power companies in the four countries that cooperate in the Nordic power system

The legal and political system

Sweden has an energy policy with certain laws and regulations and also an electricity law that defines the rules and the framework for the electric power market.

There are also international rules, regulations, and agreements. On the global level, there are targets for greenhouse gas (GHG) emission reductions, with recommendations or binding targets for individual countries. The European Union (EU) has also formulated some targets that members have to reach, and these international agreements set a framework for the energy policy in Sweden. On the other hand, the Swedish government promotes the interests of Swedish power companies, such as the state-owned Vattenfall, in the EU negotiations.

Integration of windpower

We now return to the specific properties of windpower listed in Section 3.2. How do these properties and their impact fit into the electric power system?

Moderate size and price

The moderate size and price of wind turbines opens up the opportunity for new actors to enter the power system (economic system), given that the laws allow independent power producers to connect to the electric power grid (legal/political system), which in turn will add wind turbines to the power system (technical system).

Modularity

The modularity—which means that a windpower plant may consist of anything between one and 1,000 wind turbines—adds flexibility to the power system, with shorter lead times for development than nuclear power plants; for example, it is easier to adapt power production capacity to a changing demand. To fill this role, however, the legal/political system must enable a fast process for granting permissions to such projects.

Variability

Variability (i.e., that the power production varies with the wind) will put new demands on grid operators as well as wind turbine manufacturers to enable a high windpower penetration in the power system (technical system). With higher penetration, wind turbines have to be able to contribute to the stability of the power system and participate in power regulation: to control voltage stability and frequency. Wind turbine manufacturers have thus developed technology to accomplish such demands (technical system).

The value of wind-generated electric power can be increased with good wind prognoses, which could make it possible to predict the power production during coming hours or days with high accuracy. Prognoses will, of course, not have any impact on the production of the turbines, but it will have a considerable economic value. If predictions and actual production coincide, the penalties for deviations from "ordered" power will decrease.

Distributed generation

In the early days of windpower, when the turbines were quite small (20–250 kW), they were intended to produce local power that could be used in the close vicinity, and even today many turbines actually function in that way. This is an obvious advantage for the power system, as it will reduce the losses in the grid (the longer distance power is transmitted, the larger are the losses). As long as the maximum power from the turbines is less than the minimum load in the local grid, all power from the wind turbines will be consumed in the local area.

A factory with a large and relatively constant power demand can use all power from a wind turbine for its own use. Some factories that operate day and night all year around have their own wind turbines and feed the power directly into the factory where all the power will be used.

Local resource utilization

Wind turbines utilize local wind resources. The advantage is that no raw materials have to be transported to the site during its lifetime. This saves a lot of costs for and emissions from transport vehicles but is also an advantage for energy security.

Independence of raw materials market (and price)

As no raw materials such as coal, oil, gas, or biomass have to be used and the wind is free, the costs of production are predictable. The capital costs and cost for operation and maintenance are known in advance and regulated in contracts. This property is also an advantage for energy security.

No fuel transport and no emissions to environment

Both these properties are advantages for the environment and reduce emission of GHGs and hazardous emissions. Since reduction of GHG emissions is on the political agenda (legal/political system), these properties give politicians a good motive to introduce support to windpower (economic system), which will add more wind turbines to the power system (technical system).

Note

1 Including losses in the power grid.

Appendix C

The Nord Pool market

Electric power market

The rules of the game at the Nord Pool market are described in detail by Ivar Wangensteen from Trondheim University in *Power System Economics–The Nordic Electricity Market* (Wangensteen, 2007).

He describes the power market before and after deregulation. Its focus is on economics and describes the different models used to calculate prices on the Nord Pool spot market, balance market, and so on. The certificate system is also described and analyzed. Focus is on the Nordic power market; most cases and examples are taken from Norway, but it also covers power system economics in general.

By deregulation, electric power as a physical entity and as an economic commodity were separated. Electric power was to be treated as any other commodity and prices set on a marketplace reflecting the balance between supply and demand.

However, Wangensteen points out that electricity has some features that differ from most other commodities:

- *Real time balance:* Electricity is produced and consumed simultaneously and continuously; at all times there must be an instant balance between supply and consumption.
- *Not storable:* Electricity cannot be stored in significant quantities in an economic manner.[1]
- *Consumption variability:*The consumption varies, with patterns for day/night, weekly, and annually.
- *Production variability:* The production of some renewable power plants varies as well. Windpower varies with the wind speed and the production potential of hydro power with annual and seasonal perspiration.
- *Non-traceability:* There is no way to track a unit of electric power back to a specific power plant.
- *Customer dependency:* Electric power is essential to households, companies, and most other activities. It is essential and cannot be disconnected without grave consequences and huge economic losses.

- *Overall responsibility:* In all power systems, there must be a transmission system operator who has the overall responsibility for the electric power system.
- *No real-time market:* There can be no real-time market for electricity; prices are either set ex-ante or ex-post. (Wangensteen, 2007)

Wangensteen uses the classical market price model, with price on the y-axis and quantity on the x-axis. There is a supply curve and a demand curve, and the price is set where these curves meet, at the so-called *Marshallian Supply-Demand cross.*

This is called a state of simultaneous equilibrium. It represents, according to this theory, an optimal solution for society and maximizes economic efficiency and social welfare. However, this happens only when *perfect market conditions* exist.

Preconditions for perfect competition

Perfect market conditions mean that there are many independent and competing producers and that consumers have access to all information about the market (Wangensteen, 2007). To obtain a perfectly competitive market, these conditions have to be met:

- Each market participant must be too small to be able to affect the market price (i.e., all market participants must be price takers);
- all market participants must be economically rational; producers maximize profits, consumers maximize their utility;
- all market participants must have perfect knowledge about prices and so on;
- there has to be free entry to the market; and
- there are no transaction costs.

In a perfect market, the price should, according to theory, reflect the *short-term marginal cost.* This gives maximum economic surplus, if there are no external costs.

There are, however, some sources of inefficiency. The so-called x-inefficiency is internal inefficiency in the production due to outdated technology, overstaffing, and bad management. Market inefficiency means that the price does not reflect marginal cost and marginal willingness to pay (Wangensteen, 2007).

Another source of inefficiency is imperfect competition due to *market power.* If a participant on the market is large enough to affect the market price, it can increase its profits. The simplest version of imperfect competition is *monopoly.*

With a monopoly, the producers can control the price by offering different quantities to the market. The producer is then not a price taker but a price maker. The producer will regard the price as a function of the quantity produced (Wangensteen, 2007).

If there is more than one supplier on the market, there is still a possibility that some of them can affect the market price. It is described by the *oligopoly model*, developed by the French economist Cournot in 1838.

The model is based on two preconditions:

- The producers know the demand curve.
- All producers' supply is governed by a given (known) supply from other producers.

When these preconditions are fulfilled, it is easy for all producers to optimize their profits.

According to Wangensteen, the Nordic power market is an *oligopoly*. There are a few large producers, Vattenfall, Fortum, E.ON, Statkraft, and Dong, and these companies also have a crosswise ownership (i.e., they own significant shares of one another). They have a common interest to keep electric power prices high.

Consumer response

On every free market, demand declines when prices rise. However, the response from consumers (demand) may be fast or slow. The relation between demand and price is called *price elasticity*.

Price elasticity (i.e., the consumers' sensitivity to price changes) is low on the electric power market. This may vary among different consumer categories and countries. There is also a difference between short-term and long-term elasticity. Short-term price elasticity is *not possible*, as power bills are paid as monthly average prices; to reduce power consumption when hourly prices are high does not pay.

When it comes to long-term elasticity, it is difficult to get reliable data, as there are so many factors involved. There are, however, some obvious trends when it comes to the change from oil to other heat sources in the 1980s, due to rising oil prices, and during the last years the change of electric radiators to heat pumps, which reduced the demand for electric power (Wangensteen, 2007). These are examples of long-term price elasticity (i.e., a reaction to increased prices for electric power and expected higher prices in the future).

Note

1 Electric power can be stored in hydro power dams and also by pump-storage.

Appendix D

Windpower owners in Sweden

Basic owner categories

Five ownership categories are used: private, cooperative, wind limited companies, other companies, and power companies and utilities. Here the criteria that have been used are specified.

Private

Wind turbine owned by a private person.

This is often the case also when it is part of (a machine) a farm, which is operated as a private firm. Some simple organizations such as sole proprietorship (enskild firma) and general partnership (handelsbolag) are tied to the private owner (i.e., tax declaration), and these are also classified as private.

Cooperative

Real estate communes and cooperative economic associations belong to this category. It is subdivided into consumer cooperatives, where production is tied to the members' consumption, and producer cooperatives, where all power is sold.

Wind limited companies

Limited companies with the main purpose to own and produce electric power with wind turbines.

There are some exceptions: E.ON Wind AB, a subsidiary of the power company E.ON, is sorted to the category of power companies and utilities. Windpower plants owned by subsidiaries of the real estate company Wallenstam are sorted into other companies.

Other companies

Companies with a main purpose other than windpower.

Most of these are farms operated as limited companies but also real estate companies, investment companies, and so on.

Power companies and utilities

This category is divided into transnational power companies—E.ON, Vattenfall, Statkraft, Dong, Fortum—and local/regional companies and municipal utilities, which can be public or private (see List 1).

Other ownership categories

Windpower owners have been sorted into some other categories: farmers, community power (local ownership), and consumer ownership. These criteria have been used for these classifications.

Farmers

Owners in the category "private" who are farmers.

The profession of private owners is not officially registered. I have classified owners who have the wind turbines on their land as farmers and when this is obvious. However, I have not checked the profession of all private owners, so the figure may be higher (but not lower).

Also, limited companies from the category "other companies" that, according to the official company registry (bolagsverket), are in the agricultural business are classified as farms. Additionally, other companies in the farming business (agricultural machines, plant shops, etc.) are sorted into this as a sub-category.

Local ownership

Municipal and other local utilities/power companies (see List 1):

- Wind turbines owned by farmers and farming companies in the same area.
- Wind limited companies and other companies registered in the same municipality as the wind turbines are operating (see List 2).
- Windpower cooperatives with mainly local/regional members (i.e., the national cooperatives SVEF and O2 El excluded).

Consumer ownership

All consumer cooperatives belong to this category as do real estate companies and industries, which produced the power for their own consumption (see List 3).

Sources of error

The public statistics on windpower ownership are of very bad quality.[1] For many windpower plants, especially those who do not report their production to Vindstat, no owner category is specified. In the database used for the monthly and annual reports by Vindstat, many windpower plants had been sorted into the wrong categories. All entries in the database have been checked, and these errors have been corrected. There is a remaining risk that some errors were made in this process.

Another problem is that some windpower plants on line are sold to new owners, which may belong to another owner category. Some of these changes of ownership have not been registered in the database, and it has been difficult to identify such errors. There is a risk that some windpower plants have changed from one owner category to another, which will cause an error in the shares of these owner categories. I have estimated this error to be rather small.

Many windpower plants are operated by umbrella companies, which do not own any turbines. These are, however, registered as owners or as representing the owners in the statistics. This is so also in the list of registered companies in the certificate system, as the umbrella companies administer the certificates for the owners. Such umbrella companies have been classified as windpower limited companies. The actual owners can, however, be other kinds of companies, private persons, or wind cooperatives. This will give a slightly-too-large share of windpower limited companies, and vice versa. I estimate this error to be rather small and not very important. The most interesting figure is the share of utility generators versus non-utility generators. That figure should be correct.

List 1. Local/regional power companies and utilities

Billinge Energi AB
Bixia AB
Borgholm Energi
Falkenberg Energi AB
Falkenbergs Kommun
Falu Energi & Vatten
Gnosjö Energi AB
Gunnestorp Miljö Energi AB
Göteborg Energi AB
Härnösand Energi och Miljö AB
Industrikraft
Jämtkraft AB

Jönköping Energi AB
Kalmar Energi Elnät AB
Karlshamn Energi
Kraftringen Produktion AB
Lemnhult Energi AB
Lunds Energi AB
MSE AB
Norrtälje Energi
Skellefteå Kraft AB
Södra Hallands Kraftförening
Tekniska verken Produktion, Linköping
Telge Energi AB
Umeå Energi AB
Varberg Energi
Visby Energi AB
Åsele Kraft AB
Öresundskraft Produktion AB

29 companies in list

Regional windpower limited companies

MW	*Company name*	*No. of windpower plants*
13.8	Södra Vind AB	1
28.6	Dala Vind AB	5
24.0	Samkraft AB	4

Total: 66.4 MW

List 2. Local and regional wind limited companies

kW	*Company name*	*No. of windpower plants*
600	AB Bengtssons Vind och Kraft	
800	Afsåns Vind AB	
800	AK i Asker AB	
850	Astranna Vind AB	
2,000	Backgården Vind AB	
500	Billeberga Barsebäck Vindkraft AB	
1,400	Bjälbo Energi AB	2
800	Bjälbo-Skenaån Vind	

kW	*Company name*	*No. of windpower plants*
660	Bjäresjö Vind AB	
4,000	Björka Vind AB	
750	Björnvinden AB	
850	Blås ut Vind AB	
1,200	Blåshult AB	2
4,000	Brasvind AB	2
2,450	Broby Vind AB	2
850	Broddetorps Vind AB	
2,000	Brostorp Energi AB	
4,250	Bröd AB	
6,000	Buhler och Karlsson AB	
1260	Burgvind AB	2
750	Ceding Vind AB	
2,100	Dahlin Energi AB	3
12,000	Dalboslättens Vind AB	
2,000	Dalslands Vind AB	
2,000	Drakejorden Vind AB	
1,800	Ebborp Vind AB	
500	Ekologisk energi Elfred AB	
2,000	Elvingsgården Vind AB	
850	Emfi Energi AB	
600	Enerex AB	
2,850	Erikstorp Energi AB	2
10,000	Erikstorp Utveckling AB	2
850	Farmarenergi i Åtvidaberg AB	
600	Fjällvind AB	
8,000	Fjärås Vindpark AB	
4,000	FloSal Vind AB	2
800	Freja Agri AB	
850	Frittorps Vind AB	
6,000	Frotorp Vind AB	
6,000	Gansnäs Driftbolag AB	
500	Gislöv Vind AB	2
2,000	Glömstorp Energi AB	
6,000	GSG Drift AB	
850	Gullvind AB	
800	Gunnarsons Vindkraft AB	
6,000	Gunnarstorp Vind AB	

kW	*Company name*	*No. of windpower plants*
2,000	Gärdslösa Enterprise AB	
12,000	Göteneporten AB	2
6,000	Hackeryds Vind AB c/o MSE AB	
2,400	Hafra Lyckornas Vind AB	
2,000	Hellvi Vind AB	
6,000	Herrberga Vind AB	
850	HerrgårdsVind AB	
850	Hillmill AB	
660	Himna Energi AB	
800	Holvind AB	
4,000	Hyringa Vind AB	
1,000	Hyssna Vind AB	
800	Härjevadsvind AB	
9,200	Höjdvind AB	
850	Isberget Vindkraft AB	
2,000	Isgrannatorp Enterprise AB	
750	Isiegården Vind AB	
2,000	Istad Vind AB m fl	
600	Jungavind AB	
450	K/S Galtås Vindkraft Sverige	
1,000	Kadevind AB	
10,250	Karsholms Vindkraft AB	
2,000	Kellstorps Vind AB	
850	Kilabacken Vind AB	
800	Kingelstads Vind AB	
800	Klevberget Vind AB	
800	Klockrike Vind AB	
2,000	Knäpplan Vind I	
750	Knöbbens Vind AB	
600	Kornheddinge Mölla	
850	Kraftö AB	
850	KrokaVind AB	
800	Krokek Vind AB	
225	Kråge Vind AB	
2,000	Kräklingbo Vind AB	
600	Kyrkebysvind AB	
8,000	Källeberg Vind AB	
600	Kövlinge Vind AB	

kW	*Company name*	*No. of windpower plants*
2,300	Lilla Marieholms Kraft AB	
800	LOA Vind AB	
2,000	Luftbolaget i Topperyd AB	
2,000	Lumbers Vind AB	
6,000	Lysekils Energi Vind AB	
4,000	Lårstad Vind	
6,000	Längjum Kraft AB	
2,000	Lärbo SPW AB	
800	MA Vind AB	
600	Mammarps Vind AB	
500	Manslunda Nygård	
1,000	Medvind Vindkraft AB	
1,800	Mellby Kraft AB	2
850	Millingstorp Vind	
850	Millvind AB	
600	Mälarvind AB	
6,000	Neanberg Vind AB	
600	Nordanstigs Vindkraft AB	
2,300	Näsbyvind I AB	
2,300	Näsbyvind II AB	
36,000	Näsudden Väst AB	
4,000	Oberga Vind AB	
2,000	Ringsjö Vind AB	
800	Risträsk Vind AB	
850	Roxenvind AB	
850	Runnestad Vindkraft AB	
1,850	Ryda Vind AB	2
1,000	Rydsgårds Vind AB	
3,250	Ryftes En.&Förv. AB	5
850	S. Vind El I Motala AB	
2,000	Silkomhöjden Enterprise AB	
1,500	Siral Energi AB	
850	Skallmeja BAS-Vind	
850	Skallmeja SBS-Vind	
4,500	Skara Vindkraft AB	
2,500	Skeby Vind AB	3
850	Skonberga Vind AB	
6,000	Skånings Åsaka Vind AB	

kW	*Company name*	*No. of windpower plants*
800	Stakahemmets Vind AB	
800	Stenökra AB	
2,000	Storegården Vind AB	
1,200	Svinaberga Väderspänning AB	
600	Sydax AB	
850	Sydvind Energi AB	
2,910	Söderslätts Vind AB	2
1,000	Sörgårdsvind AB	
800	Torsnäs Naturel AB	
4,500	Tre Vind AB	
4,000	Tumleberg Vindenergi AB	
850	Tungelunda Vind AB	
850	Tännäs Vind AB	
850	TÖS Vind AB	
850	Valla Vind AB	
4,000	Vallerstad Vind AB	
2,000	Vara Blåst AB	
9,000	Vassmolösa Vind AB	
800	Vidslättens Vind AB	
800	Wind Energy i Brålanda AB	
2,600	Vindenergi i Badane AB	2
750	Vindkraft Gotland AB	
450	Vindkraft i Falkenberg AB	
850	Vindkraft i Varnäs AB	
15,000	Vindpark Vänern Kraft AB	
61,000	Vindproduktion i Flackarp AB	
1,200	Vindstyrka i Trelleborg AB	2
1,000	Vindudd AB	
850	Vitåkravind AB	
850	Väderstad Vind AB	
8,350	Vänervind AB	2
800	Västervind i Skallmeja AB	
2,000	Västraby Bengt Vindkraftverk AB	
4,000	Västraby Vindpark AB	2
4,000	Vävinge Vind AB c/o MSE AB	
2,000	Åsaka-Näs Vind AB	
800	Åsen Tumleberg Vindkraft	
2,675	Ölands Vind AB	4
250	ÖmVind AB	

kW	*Company name*	*No. of windpower plants*
900	Össjö Vind AB	
725	Österlenvind AB	
800	Östra Karaby Vind AB	

Total: 471,815 kW

List 3. Consumer owned windpower by other companies

AB Botkyrkabyggen
AskersundsBostäder
Bärkehus AB
IKEA
Knivsta kommun
KopparStaden AB
Mariehus AB
Mora kommun
Mörbylånga kommun
Partillebo AB
RAJO Fastighets AB
Rättviks Fastigheter AB
Sigtuna Hem AB
Staffanstorphus AB
Wallenstam Energi AB
Öckerö Bostads AB

Note

1 The quality of the data on power production and so on are very good and valuable for owners as well as for project developers.

References

Abbad, J. R. (2010). Electricity Market Participation of Wind Farms: The Success Story of the Spanish Pragmatism. *Energy Policy 38*, 3174–3179.

Ackermann, T. A. (2001). Overview of Government and Market Driven Programs for the Promotion of Renewable Power Generation. *Renewable Energy 22*, 197–204.

Ahlbom, H. & Orring, A. (2012). Vattenfalls svek om förnybar energi. *Ny Teknik*, nr 48, 28 November 2012.

Arise Windpower AB. (2013). *Arise Windpower AB*. Retrieved 17 January2013 from www.arisewindpower.se

AWEA. (2013). Leading US Wind Power Providers in 2011. www.statista.com. Retrieved 13 March 2013, from: www.statista.com/statistics/184942/

Bayar, T. (2012). Community Wind Arrives Stateside. *Renewable Energy World*. 5 July 2012.

Bell, D. G. (2005). The "Social Gap" in Wind Farm Siting Decisions: Explanations and Policy Responses. *Environmental Politics 14*(4), 460–477.

Berkes, F. C. (2001). *Linking Social-Ecological Systems*. Cambridge: Cambridge University Press.

Bhambri, A. (2011). Country Report India. In WWEA, *Wind Energy International 2011/2012*, 119–125. Bonn: WWEA.

Bijker W. E., Hughes, T. P. & Pinch, T. (1989). *Social Construction of Technological Systems*. Cambridge, MA: MIT Press.

Bixia. (2013). *Bixia*. Retrieved 24 January 2013, from Bixia: http://www.bixia.se

Blomqvist, H. (1997). *Elkraftsystem 1–2* . Stockholm: Liber.

Böhler, T. (2004). *Vindkraft, Landskap och Mening* (Windpower, Landscape and Meaning). Göteborg: Göteborg University.

Bolinger, M. (2001). *Community Wind Ownership Schemes in Europe and Their Relevance to the United States*. Berkeley, CA: Lawrence Berkeley National Laboratory.

Bolinger, M. (2011a). Financial Innovation among the Community Wind Sector in the United States. In WWEA, *Wind Energy International 2011/2012*. Bonn: World Wind Energy Association.

Bolinger, M. (2011b). *Community Wind: Once Again Pushing the Envelope of Project Finance*. Berkeley, CA: Lawrence Berkeley National Laboratories.

Bolinger, M., Wiser, R., & Darghouth, N. (2010). Preliminary Evaluation of the Section 1603 Treasury Grant Program for Renewable Power Projects in the United States. *Energy Policy*, 6804–6819.

Bruinsma, F. V. (2008). Regional Diversity of Wind Energy Initiatives: The Dutch Experience. In J. B. van den Bergh, *Managing the Transition to Renewable Energy*. Cheltenham, UK, and Northhampton, MA: Edward Elgar.

Carlman, I. (1990). *Blåsningen—Svensk Vindkraft 1973 till 1990*. Uppsala: Uppsala University.

Constant, E. W. (1989). The Social Locus of Technological Practice: Community, System or Organisations? In W. E. Bijker, T. P. Hughes & T. Pinch *Social Construction of Technological Systems*. Boston, MA: MIT Press.

Crouch, C. (2004). *Post-Democracy*. Cambridge, UK: Polity Press.

Danish Government. (2008). *Lov om fremme av vedvarende energi LOV nr 1392 af 27/12/2008*. Retrieved 18 December 2011 from retsinformation.dk: https://www.retsinformation.dk

Dansk Vindmölleförening. (2002). Hvem ejer vindmöllerne? (Who owns the wind turbines?). *Faktablad 7*. Retrieved 29 August 2011 from Dansk Vindmölleförening: www.dkvind.dk/fakta/pdf/07

Dinica, V. (2003). *Sustained Diffusion of Renewable Energy—Politically Defined Investment Contexts for the Diffusion of Renewable Electricity Technology in Spain, the Netherlands and United Kingdom*. Enschede, NL: Twente University Press .

Dinica, V. (2008). Initiating a Sustained Diffusion of Wind Power: The Role of Public–Private Partnerships in Spain. *Energy Policy 36*, 3562–3571.

Earnest, J., & Wizelius, T. (2011). *Wind Power Plants and Project Development*. New Delhi, India: PHI Learning.

Elforsk. (1992–2012). *Driftuppföljning av vindkraftverk, årsrapporter*. Stockholm: Elforsk.

Emerging Energy Research. (2011). *Wind*. Retrieved 29 August 2011 from www.emerging-energy.com: http://www.emerging-energy.com/Research-Focus/Wind.aspx

Energimarknadsinspektionen. (2009). *Energimarknadsinspektionens rapport enligt EG:s direktiv för de inre marknaderna för el och naturgas. EI R2009:08*. Eskilstuna: Energimarknadsinspektionen.

Energimyndigheten. (2010). *Energy in Sweden—Facts and Figures 2010. Report ET2010:46*. Eskilstuna: Energimyndigheten.

Energimyndigheten. (2012). *Energiindikatorer 2012 ER2012:20*. Eskilstuna: Energimyndigheten.

Energimyndigheten. (2013a). *Driftuppföljning av vindkraftverk. Årsrapport 2012*. Eskilstuna: Energimyndigheten.

Energimyndigheten. (2013b). *Godkända anläggningar 2012-12-31*. Retrieved 3 October 2013 from www.energimyndigheten.se: http://www.energimyndigheten.se/sv/Foretag/Elcertifikat/Marknadsstatistik/

Energimyndigheten. (2013c). *Vindkraftsstatistik 2012. ES2013:01*. Eskilstuna: Energimyndigheten.

Energimyndigheten. (2013d). *Godkända anläggningar i elcertifikatsystemet*. Retrieved March 14 2013 from http://www.energimyndigheten.se/sv/Press/Nyheter/Nyhetsarkiv-2012/Godkanda-anlaggningar-i-elcertifikatsystemet2/

Energistyrelsen. (2011). *Stamdataregister for vindmöller. Oversiktstabell vindmöller pr. Ultimo juli 2011* (Master data for wind turbines end of July 2011). Retrieved 3 August 2012, from Energistyrelsen: http://www.ens.dk

E.ON. (2013). *E.ON*. Retrieved 16 January2013 from E.ON: http://www.eon.se/vind

EWEA. (2006). *Support Schemes for Renewable Energy*. Retrieved 16 April 2014 from EWEA: www.ewea.org: http://www.ewea.org

EWEA. (2010). *WindBarriers—Administrative and Grid Access Barriers to Windpower.* Brussels: EWEA.
EWEA. (2013). *Wind in Power. 2012 European Statistics.* Brussels: EWEA.
Forsström, A. (2013). Vattenfalls affärer sämst i Europa. *Dagens Nyheter*, 7–25.
Fortum. (2012). *Brighter Times Ahead.* Retrieved 18 January 2013 from Fortum: http://verkkojulkaisu.viivamedia.fi/fortumeseng/ brightertimesahead
Fortum. (2013). *Fortum.* Retrieved 16 January 2013 from Fortum: http://www.fortum.com/countries/se/om-fortum/energi-och-produktion/vindkraft/pages/default.aspx
Gan, L., Eskeland, G. S., & Kolshus, H. H. (2007). Green Electricity Market Development: Lessons from Europe and the US. *Energy Policy 35*, 144–155.
Garud, R., & Karnoe, P. (2003). Bricolage versus Breakthrough: Distributed and Embedded Agency in Technology Entrepreneurship. *Research Policy 32*, 277–300.
Gipe, P. (1995). *Wind Energy Comes of Age.* Chichester, UK, and New York: John Wiley.
Gipe, P. (2011). Locally Owned Wind Power in Germany. In WWEA, *Wind Energy International 2011/2012.* Bonn: World Wind Energy Association.
Glaser, M., Krause, G., Ratter, B., & Welp, M. (2008). Retrieved 21 January 2013, from Human-Nature-Interaction in the Anthropocene. Potential of Social-Ecological Systems Analysis.: http://www.dg-humanoekologie.de/pdf/DGH-Mitteilungen/GAIA200801_77_80.pdf
Government Report. (2001). *SOU 2001:77 Handel med elcertifikat—ett nytt sätt att främja el från förnybara energikällor.* Stockholm: Regeringskansliet.
Government Report. (2002). *Konkurrensen på elmarknaden. SOU2002:7.* Stockholm: Regeringskansliet.
Greenpeace. (2005). *Wind Force 12.* Brussels: EWEA.
GWEC. (2013). *Global Wind Statistics 2012.* Retrieved 13 March 2013 from http://www.gwec.net/global-figures/graphs/
Hardin, G. (1968). The Tragedy of the Commons. *Science 162*, 1243–1248.
Hedman, U. (2007). *Ölandsbladet.* Retrieved 17 January 2013 from Här bara pengarna "snurrar" in 2007-11-26: www.olandsbladet.se/ index.php?placid=16&placid_text=3&id=6728
Hela Sverige Ska Leva. (2012). Retrieved 26, Janaury2013 from Vindens återbäring: www.helasverige.se/fileadmin/user_upload/HSSL_Kansli/Projekt/Vindkraft/Vindens_aaterbaering_2.pdf
Hills, R. L. (1996). *Power from Wind.* Cambridge: Cambridge University Press.
Högselius, P., & Kaijser, A. (2007). *När folkhemselen blev internationell* (When Public Electricity Turned International). Stockholm: SNS förlag.
Holttinen, H. (2004). *The Impact of Large Scale Wind Power Production on the Nordic Electricity System.* Helsinki: VTT Publications, 554 pp..
Hughes, T. P. (1989). The Evolution of Large Technological Systems. In W. E Bijker, T. P. Hughes & T. Pinch, *Social Construction of Technological Systems.* Boston, MA: MIT Press.
Hughes, T. P. (1993). *Networks of Power: Electrification in Western Society, 1880–1930.* Baltimore, MD: Johns Hopkins University Press.
Hvelpelund, F. (2006). Renewable energy and the need for local energy markets. *Energy 31*(13), 1957–1966.
Hvelplund, F. (2011). Country Report Europe: Denmark. In WWEA, *Wind Energy International 2011/2012*, 204–207. Bonn: WWEA.
IKEA. (2011). *www.ikea.com.* Retrieved 14 February 2012 from www.ikea.com/se/sv/about_ikea/.../IKEA_satsar_pa_vindkraft_

IKEA. (2012). *Sustainability Report 2011*. Retrieved 17 January 2013, from http://www.ikea.com/ms/sv_SE/about_ikea/facts_and_figures/sustainability_report/sustainability_report_2011.html

Jacobsson, G., & Wizelius, T. (1972). *Ack, Värmeland du sköna—avfolkning och bolagsvälde i norra Värmland*. Stockholm: Prisma/Verdandi debatt.

Jacobsson, S., & Johnson, A. (2003). The Development of a Growth Industry—The Wind Turbine Industry in Germany, Holland and Sweden. In S. Metcalfe & U. Cantner (eds), *Change, Transformation and Development*. Heidelberg: Springer.

Jämtkraft. (2012). Vår verksamhet Retrieved 18 January 2013 from Vår verksamhet—Historik: www.jamtkraft.se/Om-Jamtkraft/Var-verksamhet1/

JP Vind. (2012). Aktuella vindkraftprojekt. Retrieved 18 January 2013 from www.jpvind.se/projekt

Kaijser, A. M. (1988). *Att välja riktning: Villkor för ny energiteknik* . Stockholm: Allmänna förlaget.

Kaintz Rognerud, K. (2010). Misslyckad satsning. *Dagens Nyheter*, 16 March 2010.

Kamp, L. M. (2004). Notions on Learning Applied to Wind Turbine Development in the Netherlands and Denmark. *Energy Policy 32*, 1624–1637.

Karlberg, L. (2009). KVA-professorer varnar för vindkraftsoptimism. *Ny Teknik*, 14 September 2009.

Karlsson, L.-I. (2011a). Systemet bäddar för dyr el. *Dagens Nyheter* 3 November 2011.

Karlsson, L.-I. (2011b). Så gör eljättarna sina drömvinster. *Dagens Nyheter* 22 February 2011

Karnoe, P., & Buchhorn, A. (2008). Denmark: Path-Creation Dynamics and Winds of Change. In W. A. Lafferty, *Promoting Sustainable Electricity in Europe*. Cheltenham, UK, and Northhampton, MA: Edward Elgar.

Klefbom, E. (2009). "Halva satsningen på vindkraft räcker." *Miljöaktuellt* 29 September 2009.

Lafferty, W., & Ruud, A. (eds) (2008). *Promoting Sustainable Electricity in Europe*. Cheltenham, UK, and Northhampton, MA: Edward Elgar.

Larsmo, O. (2007). När elen såldes ut. *Dagens Nyheter* 31 July 2007. Retrieved 17 February 2014 from http://www.dn.se/kultur-noje/kulturdebatt/nar-elen-saldes-ut/

Larsson, S. (2009). Market Director, Kalmar Energi. (23 October interviewed by T. Wizelius).

Lindahl, F. (2012). *Oppetbrev till regeringen—Ökad ambitionsnivå i elecertikatsysteme!* Falkenberg: Svensk Vindkraftförening.

Lund, H. (2010). *Renewable Energy Systems*. Amsterdam: Elsevier Academic Press.

Manwell, J. F., McGowan, J. G., & Rogers, A. L. (2002). *Wind Energy Explained—Theory, Design and Application*. Chichester: John Wiley and Sons.

Marchisio, A. (2011). Country Report Europe: Italy. In WWEA, *Wind Energy International 2011/2012*, 230–233. Bonn: WWEA.

Martin, S. (2011). The Importance of Community Power in Sustainable Energy Planning. In WWEA, *Wind Energy International 2011/2012*. Bonn: World Wind Energy Association.

Maruyama, Y., Nishikido, M., & Iida, T. (2007). The Rise of Community Wind Power in Japan: Enhanced Acceptance through Social Innovation. *Energy Policy 35*, 2761–2769.

Mauritzen, J. (2011). What Happens When It's Windy in Denmark? An Empirical Analysis of Wind Power on Price Variability in the Nordic Electricity Market. Stockholm: Research Institute of Industrial Economics-IFN.

May, H. (2008). The Global Power Station. *New Energy 3*, 08.

Mendonca, M. (2007). *Feed-in Tariffs—Accelerating the Deployment of Renewable Energy*. London: Earthscan.
Nätverket för vindbruk. (2012). *Nätverket för vindbruk*. Retrieved from www.natverketforvindbruk.se
Nelson, V. (2009). *Wind Energy Renewable Energy and the Environment*. Boca Raton, Fl: CRC Press.
Neurath, C. (2010). Elbörsen är en Kalle Anka-börs. *Svenska Dagbladet* 13 January 2010.
Nissen, P.-O. (2003). *Poul la Cour og vindmöllerne* . Lyngby: Polyteknisk förlag.
O2. (2012). *O2 får rekordorder från IKEA*. Retrieved 17 January 2013 from http://www.o2.se/press/pressreleaser/?release=678969
O2. (2012). O2 El Ekonomisk Förening. Sveriges största vindkraftskooperativ. Retrieved 15 December 2012 from http://www.o2.se/vindkraft/privat/vindkraftskooperativ/
Ostrom, E. (1990). *Governing the Commons—The Evolution of Institutions for Collective Action*. Cambridge: Cambridge University Press.
Ostrom, E. (2005). *Understanding Institutional Diversity*. Princeton, NJ: Princeton University Press.
Ostrom, E. (2007). *Sustainable Social-Ecological Systems: An Impossibility?* Retrieved 21 January 2013 from Indiana University: http://www.indiana.edu/~workshop/publications/materials/conference_papers/W07-2_Ostrom_DLC.pdf
Pengfei, S. (2011). Country Report: China, in WWEA, *Wind Energy International 2011/2012*. Bonn: WWEA.
Poteete, A. R., Janssen, M. A., & Ostrom, E. (2010). *Working Together—Collective Action, the Commons, and Multiple Methods in Practice*. Princeton, NJ: Princeton University Press.
PÖYRU. (2010). *Wind Energy & Electricity Prices—Exproring the "Merit Order" Effect*. Brussels: EWEA.
Puig I Boix, J. (2011). Community Power: The Right of People to Utilize Renewable Energy. In WWEA, *Wind Energy International 2011/2012*. Bonn: World Wind Energy Association.
Quilter, J. (2011). Sinovel to develop a 10 MW turbine. *Windpower Monthly*, 02.
Rabbalshede Kraft AB. (2013). Retrieved 17 January 2013, from www.rabbalshedekraft.se/en/About-us/
Redlinger, R. Y., Andersen, P. D., & Morthorst, P. E. (2002). *Wind Energy in the 21st Century. Economics, Policy, Tchnology and the Changing Electricity Industry*. Basingstoke: Palgrave.
Redman, C., Grove, M. J., & Kuby, L. (2004). Integrating Social Science into the Long Term Ecological Research Network: Social Dimensions of Ecological Change and Ecological Dimensions of Social Change. *Ecosystems 7(2)*, 161–171.
Rönnborg, P. (2009). *Det där ordnar marknaden*. Gothenburg: Handelshögskolan vid Göteborgs Universitet.
Schroth, G. (2011). Country Report Europe: Germany. In WWEA, *Wind Energy International 2011/2012*, 2117–2120. Bonn: WWEA.
SEAS. (2002). Offshore Projects Denmark. (T. Wizelius, interviewer)
Setterbom, J. (2010). Elbolagens samarbete för räkningarna dyrare. *Dagens Nyheter* 15 July 2010.
Sigtuna. (2012). Sigtuna kommun investerar i ett tredje vindkraftverk. Retrieved 14 May 2013 from http://www.sigtuna.se/sv/Pressmeddelanden1/Sigtuna-kommun-investerar-i-ett-tredje-vindkraftverk/

Sigtunahem. (2013). Sigtunahem köper vindkraftverk. Retrieved 16 February from http://www.sigtunahem.se/upload/Glimten/Glimten%202008/Glimten%20 2008%20nr%204.pdf

Sirland, T. (2004. Personal communication, 6 October.

Skatteverket. (2008). Skatteverkets ställningstaganden. Retrieved from 16 January 2013 from www.skatteverket.se/rattsinformation/stallningstaganden/arkiv/ ar/2008/stallningstaganden2008/13138074408111.5.121b82f011a7417 2e5880005537.html

Skatteverket. (2008). Kooperativa föreningars rabatter till medlemmar. Dnr 131 380744-08/111. Stockholm: Skatteverket.

Skellefteå Kraft. (2013). Elproduktion med stark regional förankring. Retrieved 18 January 2013 from www.skekraft.se/default.aspx?di=15674

Slitevind. (2012). Om oss. Retrieved 14 February 2012, from http:// www.slitevind. se /omoss.html

Sperling, K. (2007). *Large-Scale Wind Power in Denmark—New Perspectives on Ownership*. Aalborg: Aalborg University.

Statkraft. (2013). *Statkraft*. Retrieved 13 January 2013 from Statkraft: www.statkraft.com

Statkraft SCA Vind AB (2013). Välkommen till SCA Vind AB. Retrieved 17 February 2014 from http://www.vindkraftnorr.se/

Svensk Vindenergi. (2013). *På väg mot ett förnybart elsystem—möjligheter till 2030*. Stockholm: Svensk Vindenergi.

Svevind. (2013). Markbygden. Retrieved 16 February 2014 from http://www. svevind.se/Projects/Project.aspx?projectID=1&lang=en-US

Swedish Energy Agency. (2010). *The Electricity Certificate System, 2009*. ET2009:40. Eskilstuna: Swedish Energy Agency.

Swedish Government. (1991). *Prop 1990/91:88 Om energipolitiken*. Stockholm: Swedish Government.

Swedish Government. (1997). *En uthållig energiförsörjning prop 1996/97:84*. Stockholm: Ministry of Industry.

Swedish Government. (2000). *Ekonomiska förutsättningar för för elproduktion från förnybara energikällor prop 1999/2000:134*. Stockholm: Ministry of Industry.

Swedish Government. (2006). *Prop 2005/06:143 Miljövänlig el med vindkraft-åtgärder för ett livskraftigt vindbruk*. Stockholm: Swedish Government.

Söder, L. (1997). *Vindkraftens effektvärde*. Elforsk rapport 97:27. Stockholm: Elforsk.

Södra. (2013). *Södra*. Retrieved 16 January 2013 from Södra: http://www.sodra. com/sv/

Sörgard, I. (2009). Gefle Dagblad. Retrieved14 February 2012, from http://gd.se/ nyheter/alvkarleby/1.1312064-forsta-av-fem-vindkraftverk-pa-plats

Szarka, J. (2006). Wind Power, Policy Learning and Paradigm Change. *Energy Policy 34*, 3041–3048.

Telge Energi. (2013). Telge Energi. Retrieved 25 January 2013 from http://www. telgeenergi.se

Tengstrand, T. (2013). Bixia ProWin AB. (T. Wizelius, interview, 25 January)

TT-AFP. (2011). Tysk kärnkraft avvecklad till 2022. *Dagens Nyheter*, 30 May 2011.

van den Bergh, J., & Oosterhuis, F. (2008). An Evolutionary-Economic Analysis Of Energy Transitions. In J. B. van den Bergh & F. Bruinsmaa, *Managing the Transition to Renewable Energy*. Cheltenham, UK, and Northhampton, MA: Edward Elgar.

Vattenfall AB. (2012). *Vattenfall Bolagsordning*. Retrieved 23 July 2013 from http:// corporate.vattenfall.se/sv/file/Bolagsordning_2012-11-28_11977119.pdf

Vattenfall. (2013). *Vattenfall*. Retrieved 16 January 2013, from Våra vindkraftverk: http://www.vattenfall.se/sv/vara-vindkraftverk.htm

Veilleux, N. & Hamrick, W. (2011). Country Report: United States of America. In WWEA, *Wind Energy International 2011/2012*, 101–109. Bonn: WWEA.

Verbong, G. & Geels, F. (2008). Barriers and Options for Future Energy Transitions: Lessons from a Historical Analysis of the Dutch Electric System. In J. van den Bergh & F. Bruinsmaa, *Managing the Transition to Renewable Energy*. Cheltenham, UK, and Northhampton, MA: Edward Elgar.

Wallenstam. (2013). *Wallenstam*. Retrieved 17 January 2013 from www.wallenstam.se/om/vindkraft/vindfilosofi

Wallenstam. (2014). Vår vindfilosofi. Retrived 17 February 2014 from http://www.wallenstam.se/om/gronel/vindfilosofi/

Wangensteen, I. (2007). *Power System Economics—The Nordic Electricity Market*. Trondheim: Tapir Academic Press.

Westberg, Å. (2013). CEO SCA Energy AB. (T. Wizelius, interview, 25 January)

Wizelius, T. (1993). *Vind, del 2*. Stockholm: Larssons förlag.

Wizelius, T. (1994). *Vind, del 3*. Stockholm: Larssons förlag.

Wizelius, T. (1999). *Vindkraft en ny folkrörelse* (Windpower a New Peoples Movement). Eskilstuna: Energimyndigheten.

Wizelius, T. (2003/2007). *Vindkraft i teori och praktik*. Lund: Studentlitteratur.

Wizelius, T. (2007). *Developing Wind Power Projects*. London, UK: Earthscan.

Wizelius, T. (2009/2012). *Vindkraft på lantbruk—handbok för lantbrukare* (Windpower on Farms—Manual for Farmers). Stockholm: Vindform förlag.

Wizelius, T. (2010/2012a). *Vindkraft tillsammans—handbook för vindkooperativ* (Windpower Together—Manual for Windpower Cooperatives). Stockholm: Vindform förlag.

Wizelius, T. (2010/2012b). *Lokalt ägd vindkraft— handbook för kommuner* (Locally Owned Windpower—Manual for Municipalities). Stockholm: Vindform förlag.

Wizelius, T. (2012). Design and Implementation of a Wind Power Project. In A. Sayigh, *Comprehensive Renewable Energy (ENG)*. Oxford: Elsevier.

Wizelius, T. & Earnest, J. (2009). *Windpower in China*. Beijing: Cepri.

Wizelius, T. & Karlsson, O. (1992). *Vind, del 1*. Stockholm: Larssons förlag.

Wolsink, M. (2006). Invalid Theory Impedes Our Understanding: A Critique on the Persistence of the Language of NIMBY. *Transactions of the Institute of British Geographers 31*(1), 85–91.

Wolsink, M. (2007a). Planning of Renewables Schemes: Deliberate and Fair Decision-Making on Landscape Issues Instead of Reproachful Accusations of Non-Cooperation. *Energy Policy 35*, 2692–2704.

Wolsink, M. (2007b). Wind Power Implementation: The Nature of Public Attitudes: Equity and Fairness Instead of "Backyard" Motives. *Renewable & Sustainable Energy Reviews 11*, 1188–1207.

World Wind Energy Association. (2011). *Wind Energy International 2011/12*. Bonn: WWEA.

WWEA. (2011). World Wind Energy Report 2010. In WWEA, *Wind Energy International 2011/2012*, 13–25. Bonn: WWEA.

WWEA. (2013). *World Wind Energy Report 2012*. Bonn: World Wind Energy Association.

Index